1 CONTENT

1   Content 1

2   Introduction.. 5

3   Overview of Program Structure and Design........... 11

   3.1   Display of Toolbars and Tabs 13

      3.1.1   File Tab... 13

      3.1.2   Task Tab.. 15

      3.1.3   Resource Tab... 15

      3.1.4   Report Tab... 17

      3.1.5   Project Tab.. 17

      3.1.6   View Tab... 19

      3.1.7   Format Tab... 19

4   Setting Up a New Project 21

5   Task Scheduling 27

   5.1   Manual Scheduling/Automatic Scheduling................. 29

   5.2   Entering Tasks 33

   5.3   Dependencies 35

      5.3.1   Characteristics of Linking 39

   5.4   Structuring Tasks 41

      5.4.1   Creating a Summary Task at the Beginning of the Project...................... 41

      5.4.2   Scheduling Top-Down Summary task.................. 47

   5.5   Task Constraints 49

      5.5.1   Specifying a Deadline 49

   5.6   Task Notes.. 53

   5.7   Scheduling/Creating Milestones 55

   5.8   Timeline 57

   5.9   Critical Path.. 61

      5.9.1   Slack Times 65

6   Tables 69

   6.1   Standard Tables.. 69

   6.2   More Tables.. 71

7   Scheduling Resources 73

    7.1   Scheduling Resource Usage 73

    7.2   Inserting and Managing Project Based Resources 75

       7.2.1   Assigning Resources to Tasks 83

       7.2.2   Display Resource Usage 85

       7.2.3   Task Modes and Effort Tracking 87

    7.3   Team Planner 101

    7.4   Resource Leveling 103

       7.4.1   Automatic Resource Leveling 103

       7.4.2   Manual Resource Leveling 105

8   Cost Management 111

    8.1   Cost types 111

    8.2   Budget Tracking 123

9   Project/task Views 129

    9.1   Filter Functions 129

    9.2   Grouping 131

    9.3   Cell Highlighting 133

10   Project Control/Monitoring 135

    10.1   Saving Baseline 135

       10.1.1   Capturing Target (Set Baseline) 135

       10.1.2   Compare Target Values against Actual Values in a Table 137

       10.1.3   Compare Target and Actual Values Visually 139

       10.1.4   "Changing"/Clearing Baseline 141

    10.2   Project Continuation 143

    10.3   Evaluating Monitoring Information 145

11   Custom Fields 147

    11.1   Lookup Fields 147

12   Multi-Project Management 153

    12.1   Subprojects 155

    12.2   Create Resource Pool 159

    12.3   Project portfolio/Overview 161

13   Reports and Graphic Evaluations 163

   13.1   Visual Reports 163

   13.2   Overview of Visual Reports 165

       13.2.1   Category "Task Usage" 165

       13.2.2   Category "Resource Usage" 165

       13.2.3   Category "Assignment Usage" 167

       13.2.4   Categories "Task Summary & Resource Summary" 169

   13.3   Reports Directly From Microsoft Project 171

14   Attachments 175

   14.1   Working With Other Office Programs 175

   14.2   Earned Value Analysis with Microsoft Project 177

   14.3   Complete List of all available Fields in Microsoft Project 185

   14.4   Basic Settings 185

       14.4.1   Naming Standards 185

       14.4.2   Standards for Summary Tasks 185

       14.4.3   Standards for Milestones 187

       14.4.4   Standards for Tasks 187

   14.5   Book Recommendations 189

   14.6   Glossary 192

   14.7   Keyboard shortcuts for Project 2016 203

15   Index 213

NOTES, COMMENTS:

2 INTRODUCTION

The current version of Microsoft Project 2016 provides extensive project planning options also in "standalone" mode (without MS Project Server connection). The following versions of Microsoft Project are available:

Feature	Project Standard 2016	Project Professional 2016	Project Pro for Office 365
Quick start of new projects	☑	☑	☑
Project tracking for proactive detection of critical paths	☑	☑	☑
Quick generation and locating reports	☑	☑	☑
Additional apps from the Office store	☑	☑	☑
Easy resource management	☒	☑	☑
Call or instant message to team members from project (requires Lync Server)	☒	☑	☑
Collaboration with others from almost anywhere (requires SharePoint and Project Server)	☒	☑	☑
Growth opportunities	☒	☑	☑
Use of most recent version of MS Project	With Software Assurance	With Software Assurance	Auto-matically
Access from almost anywhere	☒	☒	☑

The standard version CANNOT be updated. Therefore, the insignificantly higher price of the professional version is recommended to also allow a Microsoft Project Server solution, if needed.

NOTES, COMMENTS:

This document describes the version MS Project Professional 2016 that, compared to the version MS Project 2013 and apart from a possible Microsoft Project Server connection, provides the following additional features:

- Presentation of several timelines in parallel, e.g. summary tasks and detailed tasks in two timelines

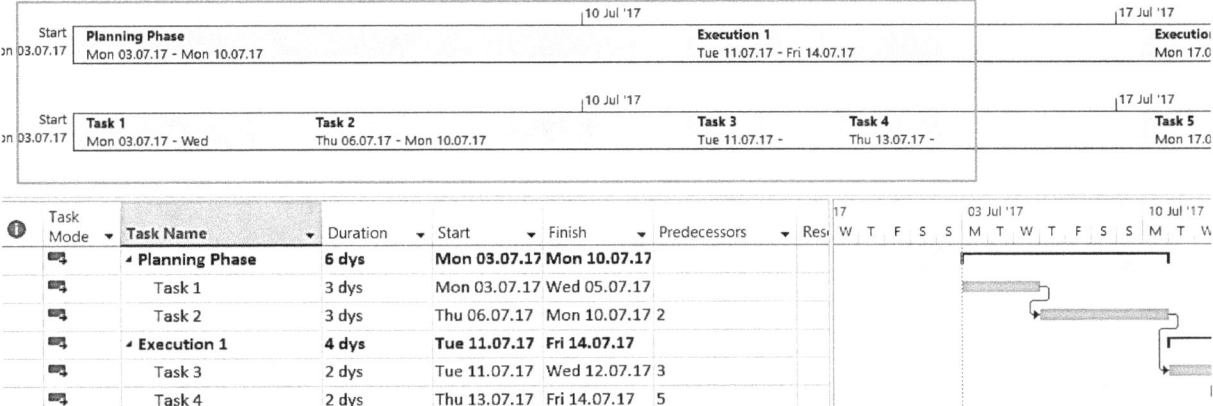

- "Tell me what you want to do". Using this search field, you are directly guided to the corresponding function/command.

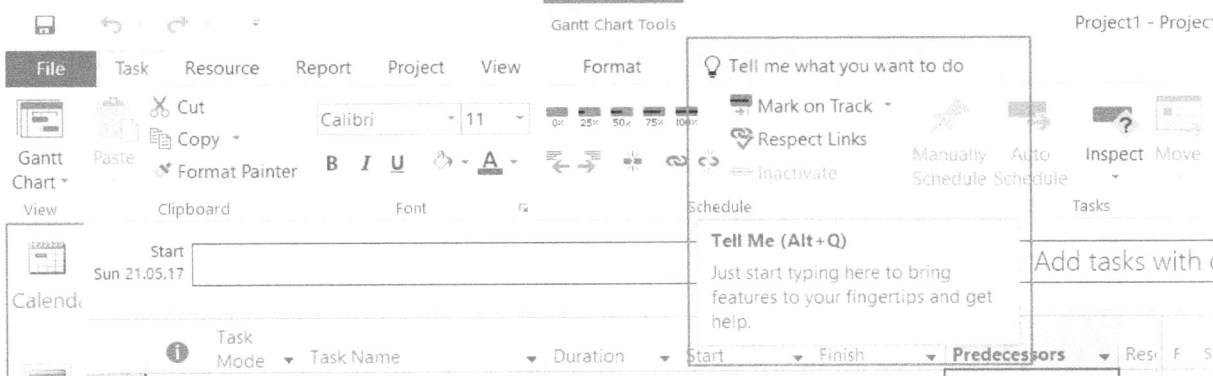

The number of modifications compared to version 2013 is relatively small. There are some changes in the current Microsoft Project Server version in the area of resource scheduling.

In principle, the interface of the current Microsoft Office version, i.e. menus and tool bars will be replaced by the toolbar/the ribbon and will be described in detail on the following pages.

NOTES, COMMENTS:

Each task of project management can be used as a support in Microsoft Project across all planning steps.

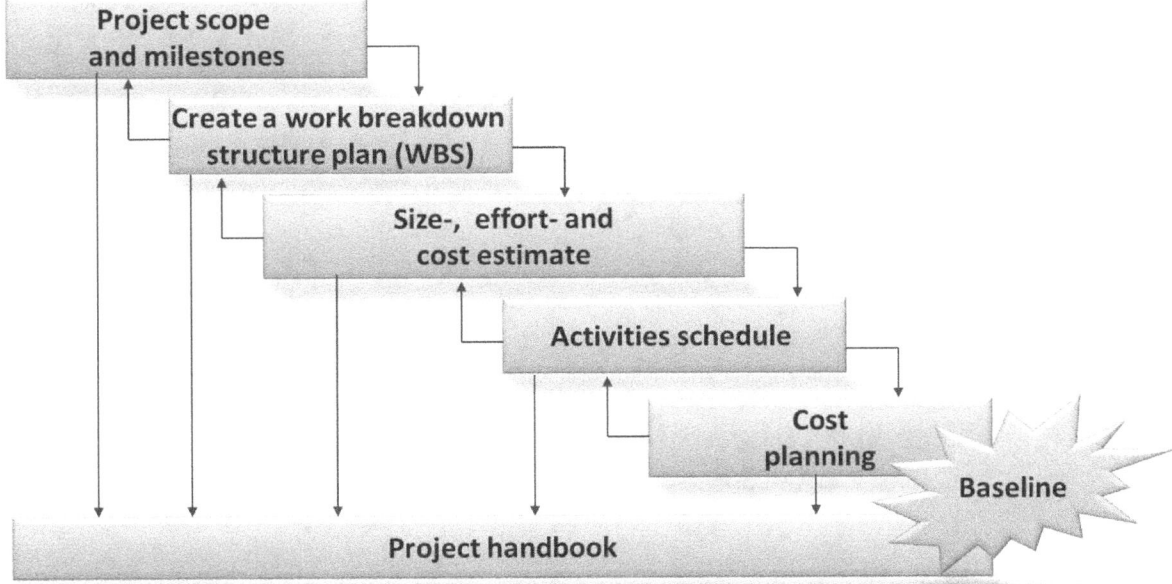

- A rough scheduling of milestones defines the entire project and its duration.

- Work packages can be structured using the outline function.

- The duration and work (resource) are determined per work package.

- The task schedule is generated by existing task relationships including predecessor and successor.

- Based on assigned resources (work and/or material), a detailed cost budget is established.

- The function "Save Baseline" allows a consistent target/actual comparison with the planning baseline.

NOTES, COMMENTS:

3 OVERVIEW OF PROGRAM STRUCTURE AND DESIGN

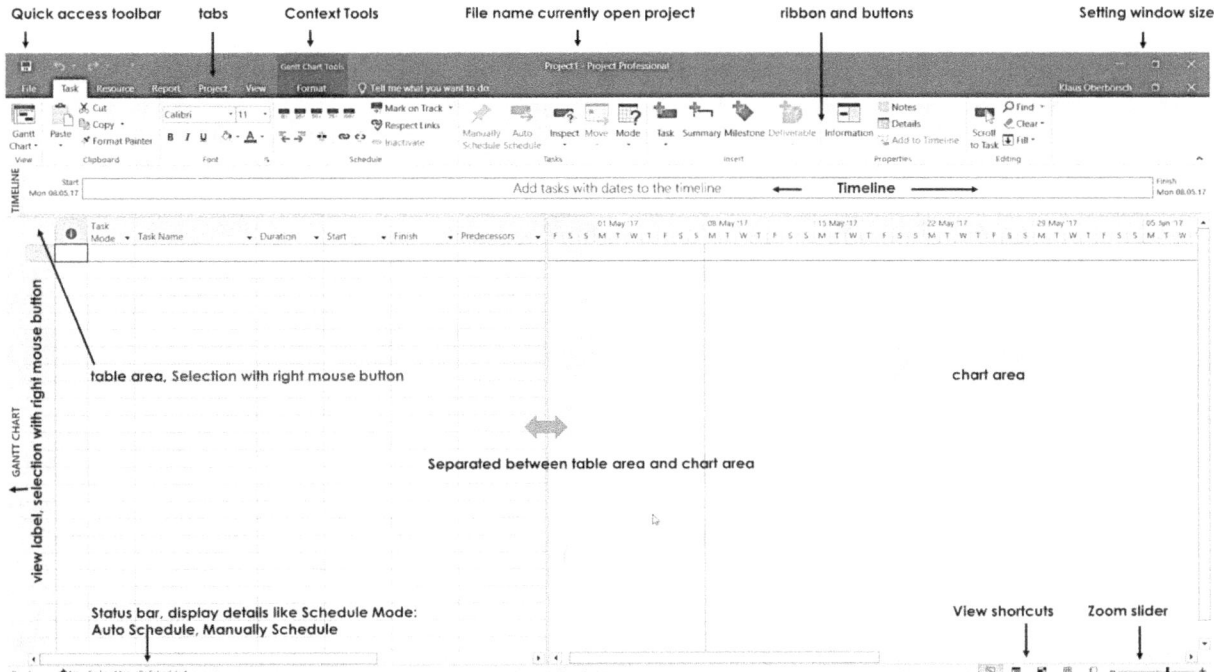

The separating line (view area) between table and diagram area can be adapted using the left mouse button. The command groups of each tab will be described below.

Important and frequently used functions can be directly accessed using the toolbar. Almost all functions can be integrated using "Quick access toolbar".

NOTES, COMMENTS:

3.1   DISPLAY OF TOOLBARS AND TABS

3.1.1   FILE TAB

The toolbars and tabs essentially show all central commands such as "Save", "Save as...", "Print" and "Options". This view also enables you to execute direct imports or exports to other programs such as e.g. Excel and PDF and others.

In the menu item "Share", you can find the direct file transfer of the current Microsoft Project file via email or release under "SharePoint Services".

The most important settings in the menu item "Options" will be described later.

NOTES, COMMENTS:

3.1.2 TASK TAB

This tab allows you to make direct settings for task scheduling. In detail:

1. View (e.g. switch windows between Gantt Chart, Network Diagram, Resource views)
2. Insert from Clipboard
3. Fonts and Format
4. Time Schedule, Degree of Completion, Deactivate Tasks
5. Tasks mode (switch from Manually Scheduled to Auto Scheduled)
6. Insert options: Insert Tasks, Summary Tasks, Milestones
7. Properties/Information
8. Editing, Scroll to Task, Find, Clear

3.1.3 RESOURCE TAB

This tab contains all functions for resource assignment and resource management. In detail:

1. Team Planner, Assign Visual Resources
2. Assignments, Assign Resources, Resource Pool
3. Insert, Add Resources
4. Properties, Resource Properties, Notes and Details
5. Leveling Options

NOTES, COMMENTS:

3.1.4 REPORT TAB

As of this version, users benefit from a comfortable reporting function that generates direct reports in Microsoft Project: You can choose from various evaluations of resources, costs, processing status etc. All reports can be customized. Furthermore, the function "Visual reports" is available generating reports in Excel or Visio.

1. Compare Projects, Compare Different Project Versions
2. View Reports, select different report types integrated into Microsoft Project
3. Visual reports, Export to Excel or Visio with set report

3.1.5 PROJECT TAB

This tab allows to set basic information. In detail:

1. Insert Subproject, Multi-Project Management
2. Properties, Project Information, Custom Fields, Links Between Projects, PSP Code, Change Working Time
3. Time schedule, Calculate Project, Set Baseline, Move Project
4. Status, Update Project
5. Spelling, Proofing

NOTES, COMMENTS:

3.1.6 VIEW TAB

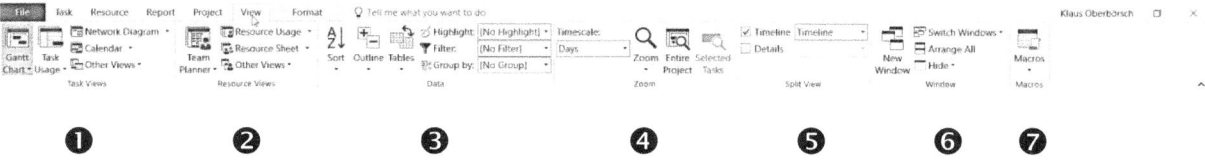

This tab enables you to select all possible views such as Network Diagram, Calendar, and Team Planner. This way, you can manage the timeline and enable filter and grouping functions. In detail:

1. E.g. Task Views, select Other Views, Gantt Chart, Task Usage
2. Resource Views, Team Planner, Resource Usage
3. Data, Sort, Filter, Group by
4. Zoom
5. Display elements, enable Timeline, select Details
6. Window
7. Macros, insert macro, play

3.1.7 FORMAT TAB

This tab allows you to format the text and the bars as well as to display and show critical tasks, slack time, outline number and project summary task.

1. Text Styles, Layout, Gridlines
2. Columns, Insert column, Custom Fields
3. Bar Styles, mark Critical Tasks, Slack
4. Gantt Chart Style, selection of predetermined color schemes
5. Show/Hide, Outline Number and Summary Task
6. Drawings, Drawing

NOTES, COMMENTS:

4   SETTING UP A NEW PROJECT

For setting up a new project, you can start or open an empty project by selecting predetermined templates for project plans.

The menu "File/New" shows all existing options.

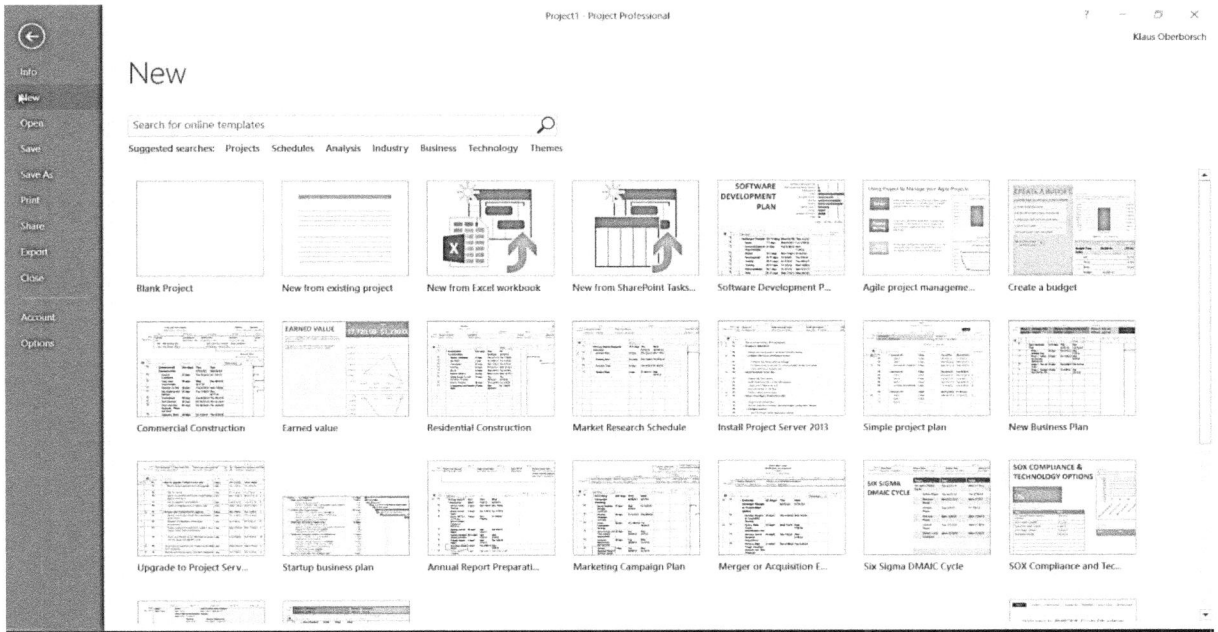

Here are predefined project plans for diverse planning projects available as well as an import from an Excel workbook or a SharePoint Tasks list. Other project plans are available using "Online Templates".

NOTES, COMMENTS:

Let's take a look at **how to create a new project**. Basic information about the project is entered under "Project/Project Information".

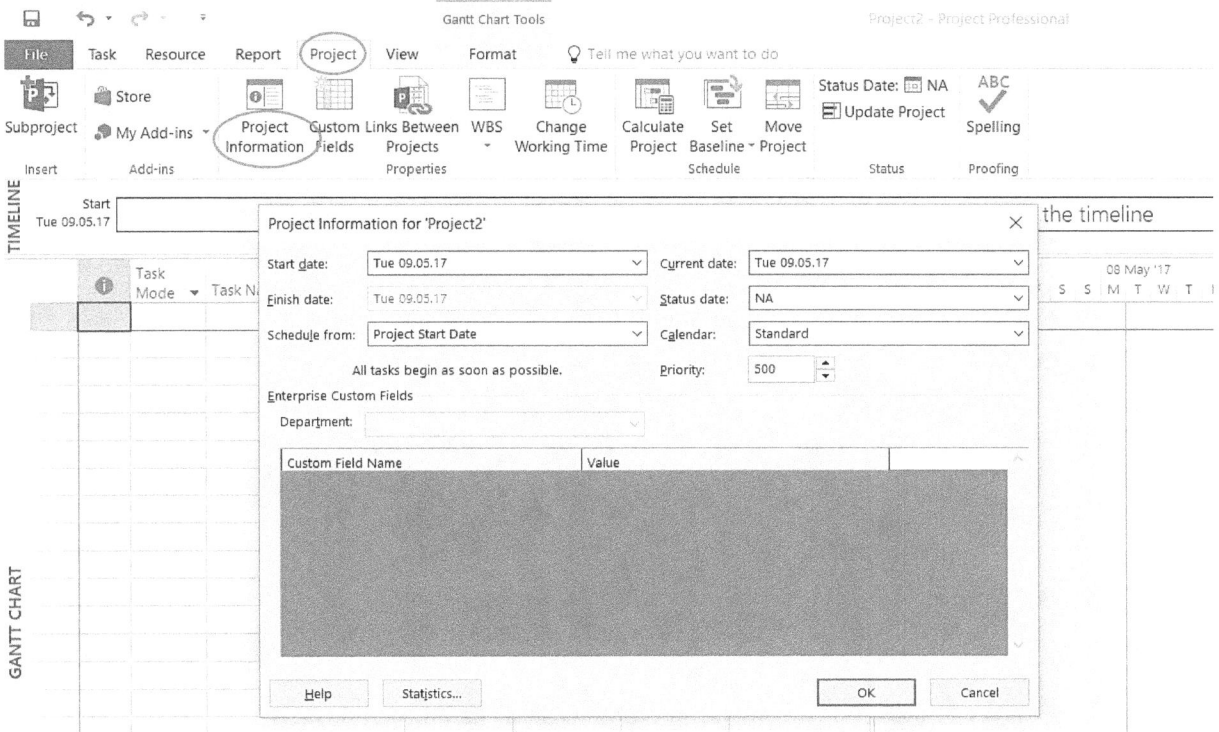

The "***Start date***" of the project is recorded in the project information window. The start date of the first task doesn't need to be identical with the start date! If tasks have already been entered, the "***Finish date***" will be calculated accordingly. The field "Finish Date" actually serves the next entry, "Schedule from:".

The selection field "***Schedule from***" serves as the basis on how to calculate the project. Here, you can select the Start date or the Finish date of a project. Both variants are also referred to as forward or backward pass.

The difference is that, in case of the forward pass, all tasks begin as soon as possible due to the set start date. If the forward pass is used, all tasks begin as late as possible. The forward pass is used if the start date of the project has been specified and you would like to calculate the end of the project (finish date) by means of Microsoft Project. The forward pass is used with projects that must be finished at a particular date (e.g. conversion to the Euro in 2000).

The field "***Current date***" shows the system date. If it doesn't correspond to the actual date or for simulations at a future date, it can be changed here.

You can determine the "***Status date***" later. The status date will then be selected to determine reports about time, costs or performance of a project instead of the current date at the selected date. As long as you haven't set a start date, the field displays "***NA***" (= not available).

NOTES, COMMENTS:

Select the standard calendar for the project in the field "**Calendar**". A standard calendar contains project-typical working times and work-free times for the project. The settings of the standard calendar should be checked prior to scheduling and can be changed later, if needed.

In the field "Priorities", you can specify the priority of this project. This information is important when using a multi-project technique, if for example subprojects are moved in dependence of the importance (priority) to relieve overallocated resources.

Furthermore, the scheduling type is displayed, in the screenshot "All tasks begin as soon as possible", and this corresponds to the standard settings. The field "**Custom fields**" is only relevant here for a solution via the Microsoft Project server.

NOTES, COMMENTS:

5 TASK SCHEDULING

After the creation of a project, the beginning of the scheduling mainly focusses on inserting and managing project tasks. In Microsoft Project, project tasks are referred to as tasks. The temporal assignment of tasks among each other is referred to as task link or relationship.

The duration of a task or duration is the pure period of time that is required to complete a task. It depends on the resources and other settings how much work can be done during this time. The duration can be specified in minutes, seconds, days, weeks or months. The set duration is used to calculate the start and finish date of the task. The work to be done can only be calculated by assigning resources!

The following units are available to specify the duration:

Unit	Meaning	Example
m	Minutes	90 min
h	Hours	36 hrs
d	Days	2 days
w	Weeks	2 wks
mo	Months	1 mo

If you use another unit other than day, you must also indicate the number of the desired unit after entry.

In Microsoft Project, there is another type of duration: the ongoing duration. Ongoing duration describes the time that is required to execute a task based on 24 hours and 7 days a week, including public holidays and work-free days. Ongoing durations are defined by using the prefix "e" during entry, meaning emin, ehrs, edays, ewks, emo. Ongoing durations can be used in production processes, e.g. with machines that run all day (and night) long.

NOTES, COMMENTS:

5.1   MANUAL SCHEDULING/AUTOMATIC SCHEDULING

Since version Microsoft Project 2010, an important modification was introduced with regards to project scheduling. By making changes to factors such as the relationships and the project calendar, the task data is no longer adapted automatically if a task will be scheduled **manually**. This means that scheduling can be executed manually such as e.g. a simple schedules display in Excel without automated routines in the background.

This setting can be made per task or for the entire project. The right mouse button on the respective row number displays the scheduling mode for the selected task in a pop-up message.

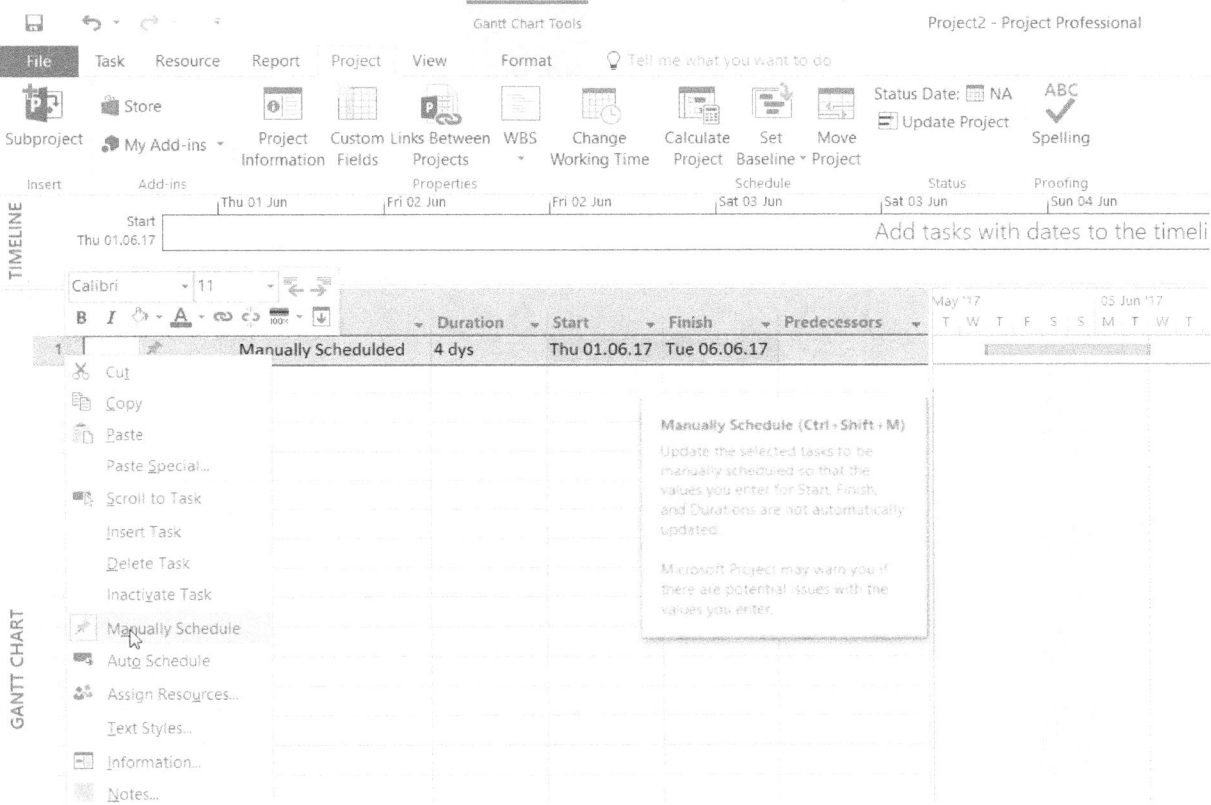

In the column "Task Mode", a pin is shown as a reference to Manually Schedule and the task bar has another format than Auto Schedule tasks. Auto schedule tasks are marked with a task bar as well as an arrow in the column "Task Mode".

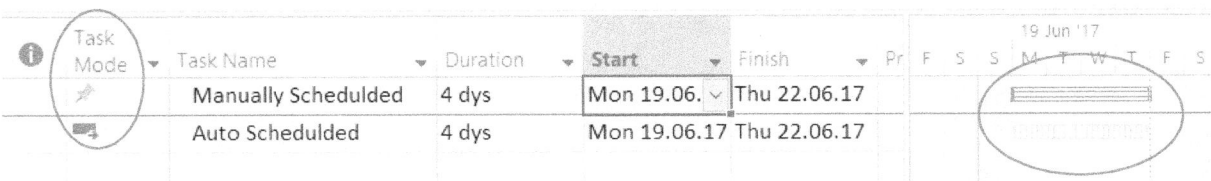

NOTES, COMMENTS:

The scheduling type, which is set for the entire project, is shown in the status bar (bottom left).

It can be changed in the dashboard under "File/Project Options/Schedule" for the current project or for all **new** projects.

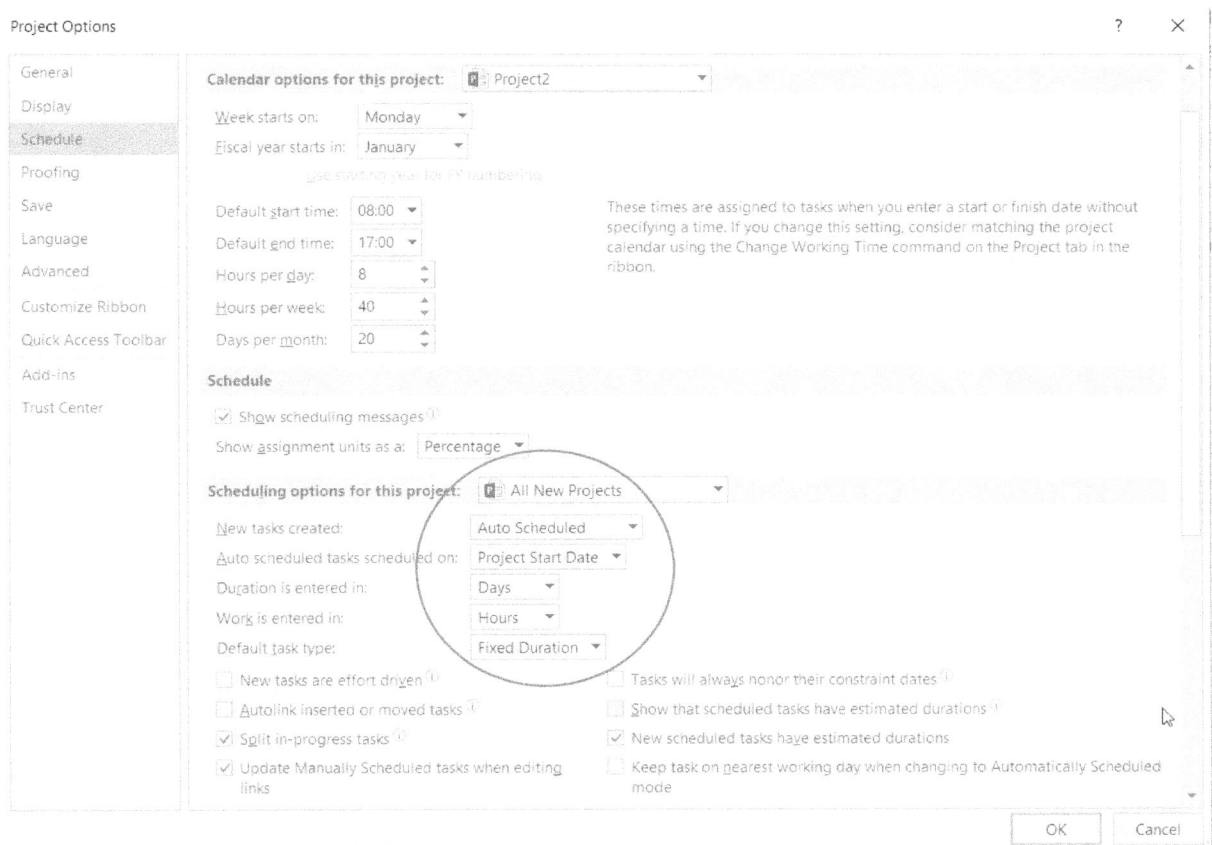

NOTES, COMMENTS:

5.2   ENTERING TASKS

Please use the view "Gantt Chart/Tables" below (default view when accessing Microsoft Project) to insert tasks. In case another view is displayed, call up the desired view through the menu "View/Gantt Chart" in combination with "Tables". The easiest way to select the table is by clicking with the right mouse button on the intersection point of the rows and columns (see mark), alternatively by using the menu item "View Tables".

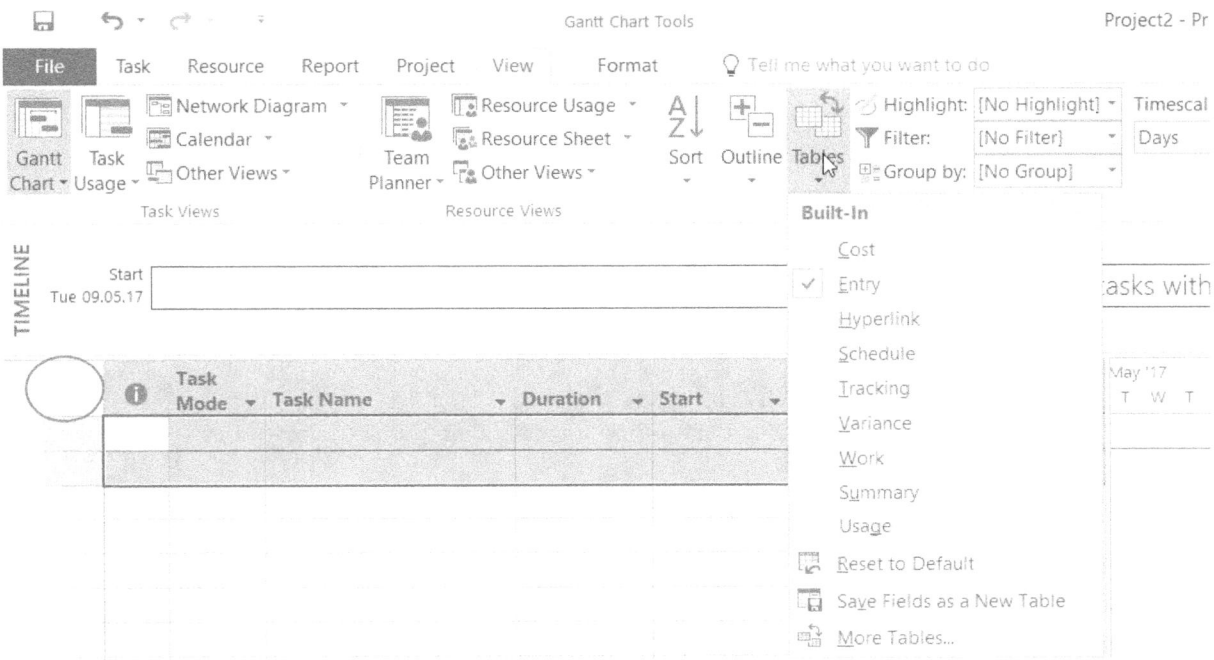

The respective task descriptions should contain meaningful names. After entering the names, you can also insert the duration or the start and finish date resulting in an automatic calculation of the duration. A bar is drawn into the calendar in line with the duration. In case you don't specify any unit, Microsoft Project either enters the duration in days by default or the most recently used unit.

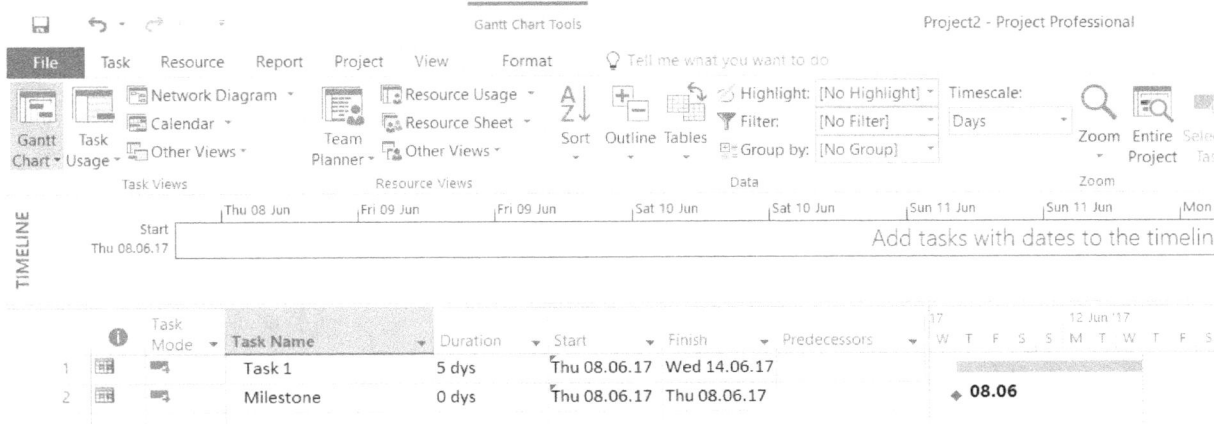

Milestones are inserted with a duration of **0 days**. This process will later be described in detail in chapter 5.7.

NOTES, COMMENTS:

5.3 DEPENDENCIES

Relationships or task links reflect dependencies between tasks. For this purpose, you set tasks in a relationship to each other.

This happens in Microsoft Project if you assign the predecessors or successors to each task that – in a particular way – are a requirement for completion.

- The task, that another task depends on, is the predecessor of this task.
- The task, whose start or finish depends on another task, is called successor.

The following task links are presentable in Microsoft Project and are also defined that way in general descriptions concerning project management:

Finish-to-start (FS)	Task B can only start once task A has finished.	
Start-to-start (SS)	Task B must start once task A has started.	
Finish-to-finish (FF)	Task B must be finished once task A has finished.	
Start-to-finish (SF)	Once task A starts, task B must be finished.	

The task links can be executed through different steps. One option is to add explicit row numbers of the respective task that visually displays the links. It's recommended to also display the column "Successor". This way, you can enter the predecessor or successor relationships optionally. Microsoft Project automatically calculates the predecessor or successor depending on your entry.

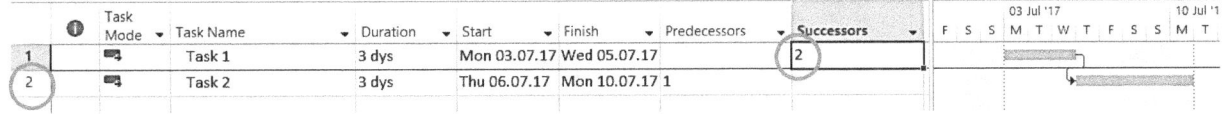

	ⓘ	Task Mode	Task Name	Duration	Start	Finish	Predecessors	Successors	03 Jul '17 F S S M T W T F S S M T	10 Jul '1
1			Task 1	3 dys	Mon 03.07.17	Wed 05.07.17		2		
2			Task 2	3 dys	Thu 06.07.17	Mon 10.07.17	1			

Entry of row numer
Predecessor **or** successor

NOTES, COMMENTS:

Or you create a link using the mouse. Simply position the cursor in the center of the task from where the successor relationship should be created. Press and hold the left mouse button. Use the cursor to drag it to the center of the task with which you want to create a finish-to-start relationship.

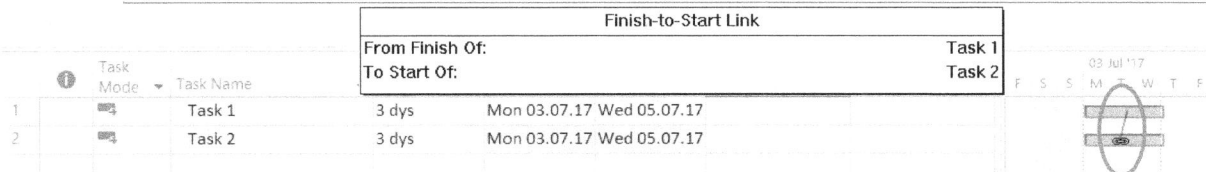

For control reasons, the desired task relationship is displayed in a window. The fields "Predecessor" and "Successor" are filled automatically.

The task relationships can also be mapped by double-clicking on the Successor task in the respective row. The window "Task Information" opens. The task number can directly be entered under the Predecessor tab. Alternatively, the respective task can be selected by using the drop-down menu "Task Name".

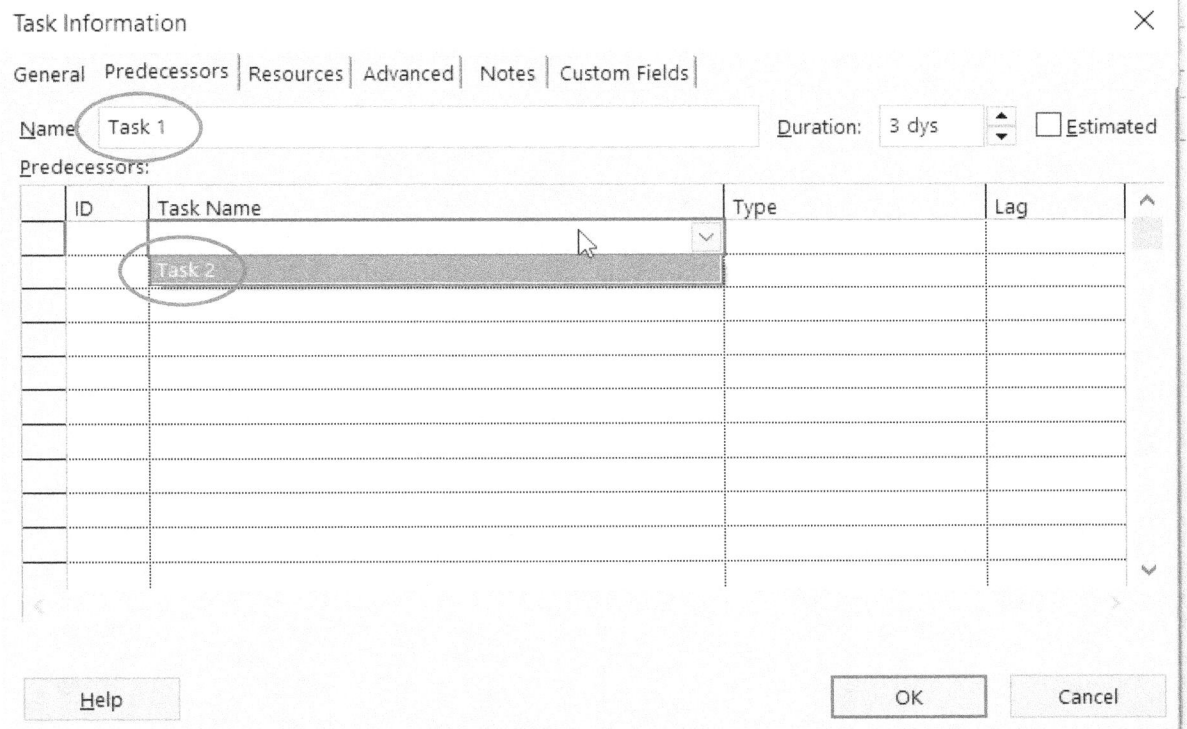

NOTES, COMMENTS:

5.3.1 CHARACTERISTICS OF LINKING

In case the standard link "Finish-start" isn't desired, you can use the tab in the task information column "Type" and "Lag" to make all settings.

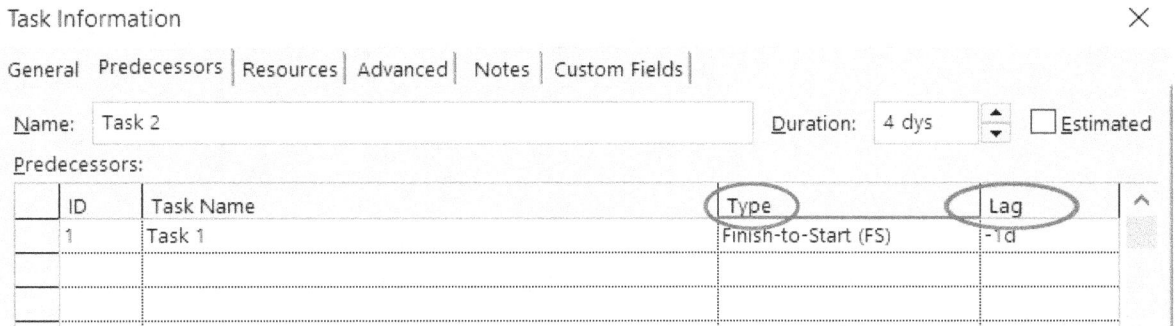

The following characteristics are possible:

Predecessor column	Function
2	The task can only start after the finish of task 2.
2;3	The task can only start after the finish of task 2 and after the finish of task 3.
2EA-1t	One day prior to the finish of task 2, the task must start. Entry in the column "Lag".
2AA	The task must start when task 2 starts.
2;3AA	The task can only start after the finish of task 2 and at the same time as the start of task 3.
5EE	The task must finish when task 5 finishes.
3AA+3t	The task must start 3 days after the start of task 3. Entry in the column "Lag".
4AA+40%;3EA+3t	As soon as 40% (and thus dynamically adapted to the actual duration of the predecessor) of task 4 have been executed, this task must start. Also, task 3 must already have been finished for 3 days. Entry in the column "Lag".

All types of task links can also be executed across different project, i.e. you can create dependencies across subprojects. Subproject 2 can only start when subproject 1 has delivered a specific service (multi-project management).

NOTES, COMMENTS:

5.4   STRUCTURING TASKS

Extensive projects consist of main tasks to which, in turn, subtasks are assigned. This results in a textual outline of the project tasks. Such an outline provides a better overview for large projects.

Microsoft Project includes an outline tool with which you can realize a textual outline from a program technology perspective. In Microsoft Project, the phases or summaries of tasks are referred to as summary tasks.

You can enter summary tasks right when first entering all tasks but you can also add it to a task list retrospectively.

5.4.1   CREATING A SUMMARY TASK AT THE BEGINNING OF THE PROJECT

The summary task should have an explicit name to better identify or assign the underlying tasks. In this example, we call the summary task "Planning Phase". The description is entered as a normal task, the duration can stay as is for now.

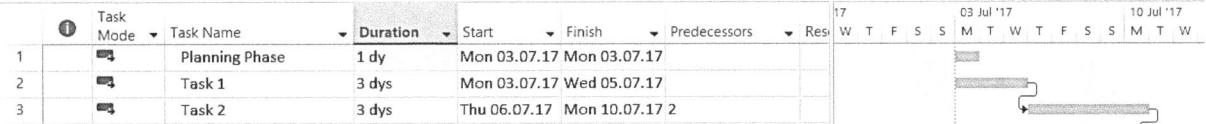

	ⓘ	Task Mode ▾	Task Name ▾	Duration ▾	Start ▾	Finish ▾	Predecessors ▾	Res
1		🔧	Planning Phase	1 dy	Mon 03.07.17	Mon 03.07.17		
2		🔧	Task 1	3 dys	Mon 03.07.17	Wed 05.07.17		
3		🔧	Task 2	3 dys	Thu 06.07.17	Mon 10.07.17	2	

Please note: All tasks must be set to "Auto Schedule". The duration of the summary task results from the longest duration of all underlying tasks.

If all tasks related to "Planning Phase have been entered, they are marked as follows: Left mouse button on the corresponding row number and mark the desired lines with the mouse button held down. Alternatively, mark each row using the CTRL button.

NOTES, COMMENTS:

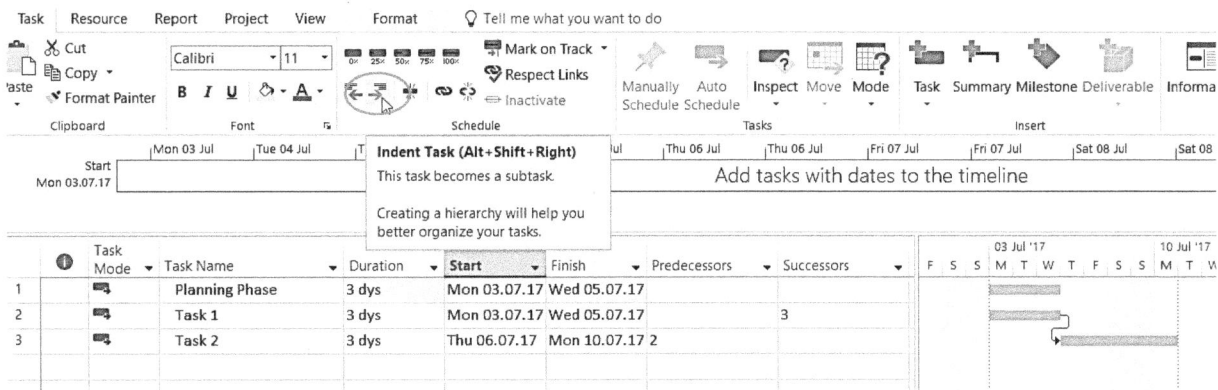

Afterwards, click the green arrow to the right under menu item "Indent Task". The marked tasks will be indented to the right, and the summary task now occupies the duration of the longest task chain within the summary task.

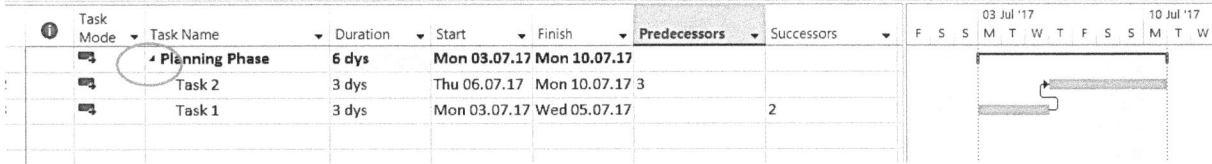

The detailed tasks can be shown and hidden with one mouse click on the small triangle in front the summary task. You can indent tasks also on other outline levels.

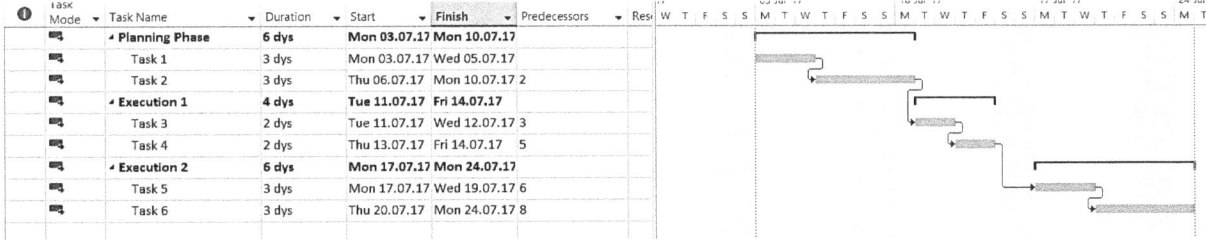

NOTES, COMMENTS:

To avoid losing the overview in case of extensive outlines, it's recommended to also display the outline number of each task, check the checkbox on the right in the menu item "Format".

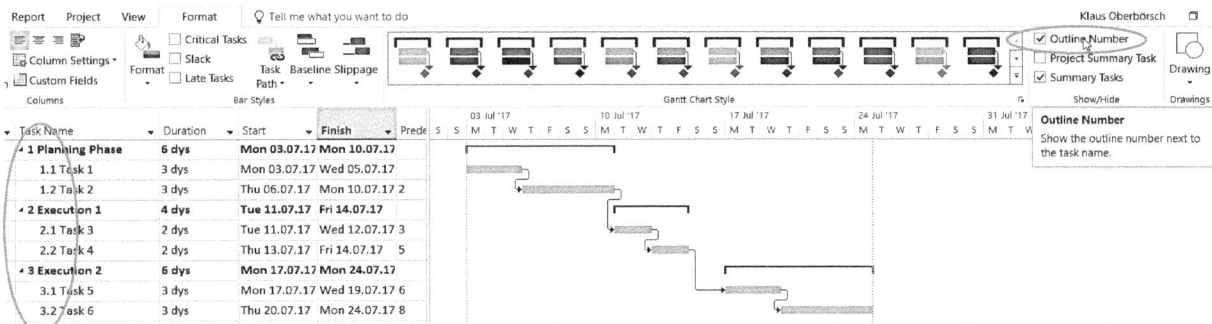

Alternatively, you can also display the outline number in a separate column. Mark the column header where the outline number should appear, right mouse click, insert column and select the field "Outline Number" or "PSP Code".

NOTES, COMMENTS:

5.4.2 SCHEDULING TOP-DOWN SUMMARY TASK

The possibilities of scheduling projects are no longer limited to solely creating subtasks and displaying them in summary tasks. Since version Microsoft Project 2010, summary tasks can be created with a date that doesn't need to be precisely identical to the data of the subtasks.

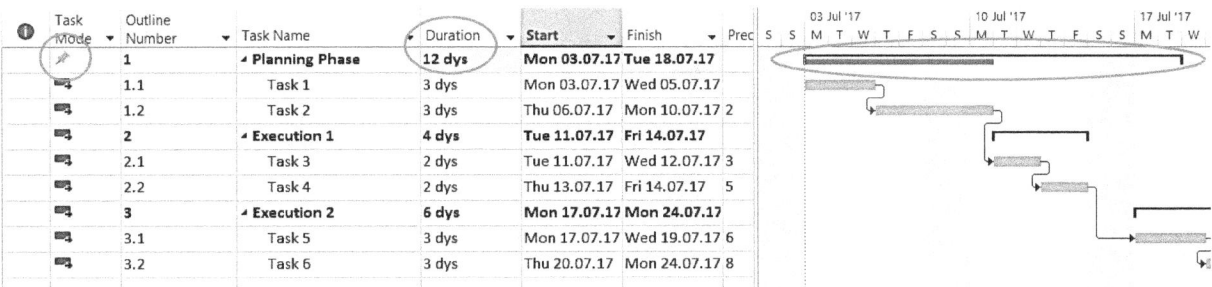

At the beginning of a planning phase, you may only have some superordinate information on the most important scopes of delivery and milestones of your project. The "rough planning" will be displayed in the summary task, but, by comparison, also the effective display of time of the related subtasks. Therefore, the task will be marked as summary task after indicating the start and finish date or the duration, most easily through the menu item "Format" and by clicking on the function "Summary Task". Please note: The task must be entered as "Manually Scheduled".

Here, you can find the added "real" duration of each task under the rough estimate of the requirements analysis. A summary task, defined this way, adapts to the underlying tasks after enabling the automatic scheduling of the effective duration.

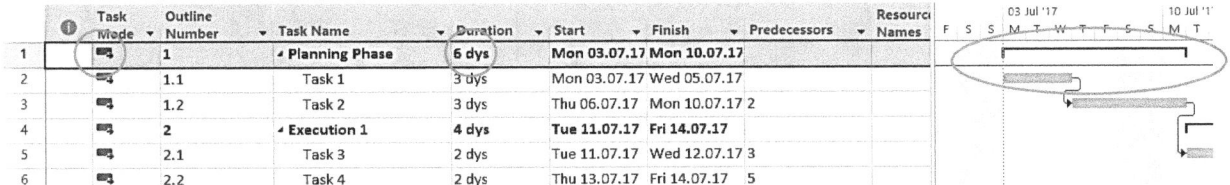

NOTES, COMMENTS:

5.5   TASK CONSTRAINTS

You can bind tasks to defined dates by means of task constraints. Such constraint dates are e.g. used in the following situations:

- In the case of milestones that should be reached at a set date.
- In the case of objective reasons for a constraint, e.g. event dates within a project that are fixed or in the case of road works that must be completed before winter.

5.5.1   SPECIFYING A DEADLINE

A visual presentation of a date as a deadline is one way of marking constraints or alerts. The **constraint** will be displayed by an icon in the indicator column.

If the task is delayed, so that the finish must be moved beyond the deadline, a note appears in the indicator column.

1. Double-click on the task or icon in the ribbon to which you would like to assign a constraint date.
2. Switch to the **Advanced** tab in the Information window of the task.
3. Select the desired constraint types from the list of "**Constraint types**".
4. Enter the date for the constraint in the field "**Deadline**".
5. Select "Must Finish On" under "**Constraint type**", click OK
6. Update the message of the "**Planning Wizard**" with "Continue"
7. Click **OK.**

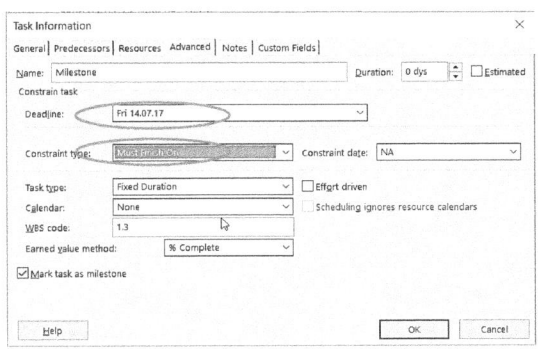
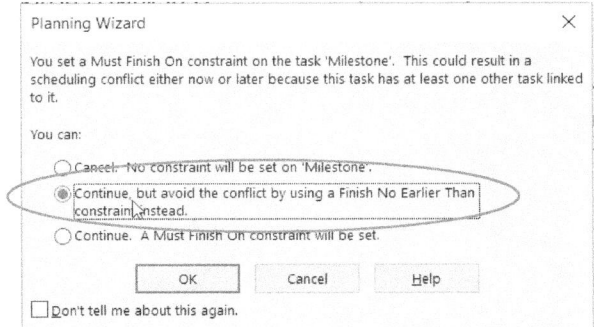

NOTES, COMMENTS:

The constraint will be displayed by an icon in the indicator column and the set due date by a green arrow at the task (here: Milestone).

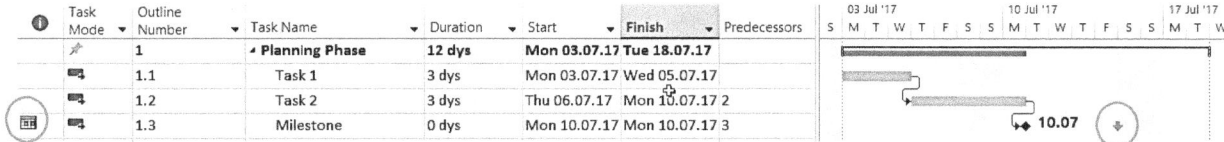

Now, extent the preceding tasks (task # 2 has extended from 3 to 8 days) so that the Milestone moves to a later date accordingly. The green arrow for marking the deadline will remain on the original date and the current deadline will be displayed to the right. Furthermore, a note will appear in the indicator column showing the due date.

NOTES, COMMENTS:

5.6 TASK NOTES

Individual notes can be added to each task and saved in the Microsoft Project file. This way, information can be inserted without any additional tools such as Word etc.

To do so, double-click on each task. Then select the tab "Notes" in the window that opens. It's also possible to link to objects from other applications (e.g. PowerPoint).

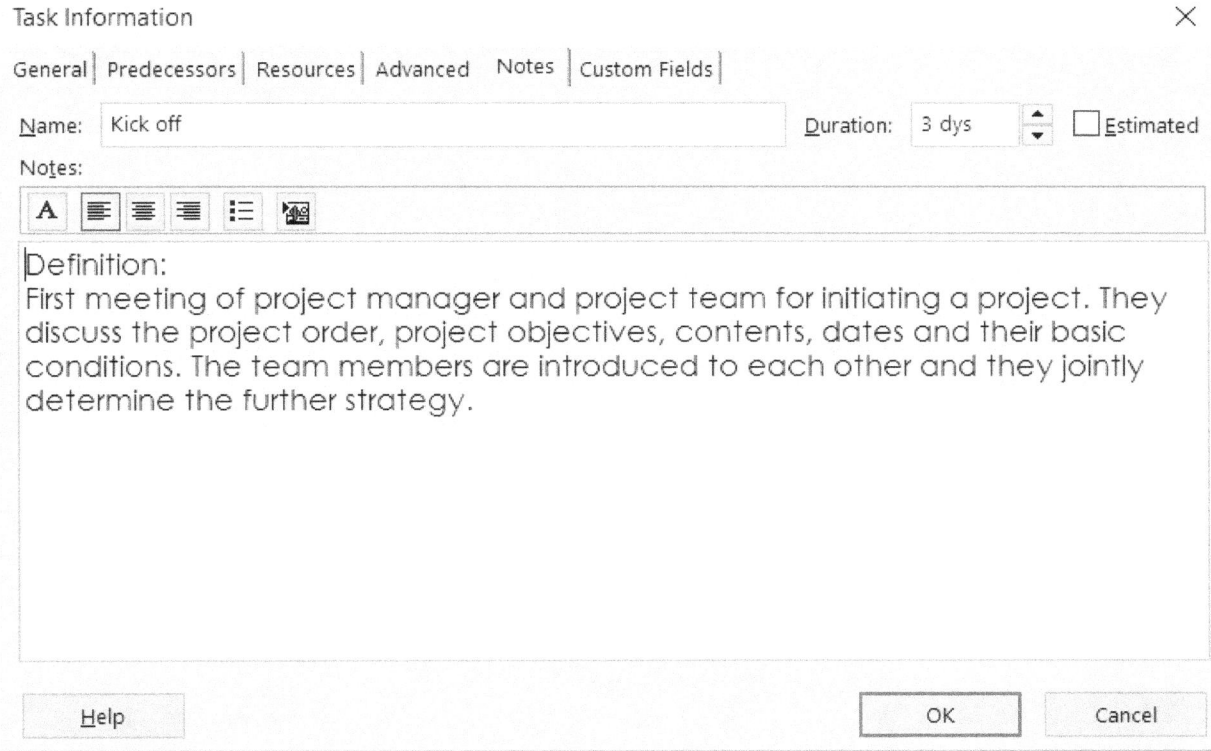

In the indicator column, an information on saved notes is shown by a post-it note. If you move the cursor over each post-it note, parts of the note are displayed. Alternatively, the field "Notes" can be inserted in each table.

However, for information that should be used later for groupings or filter function (e.g. responsible person, cost center), it's recommended to use custom fields as described under section "Custom Fields" in chapter 11.

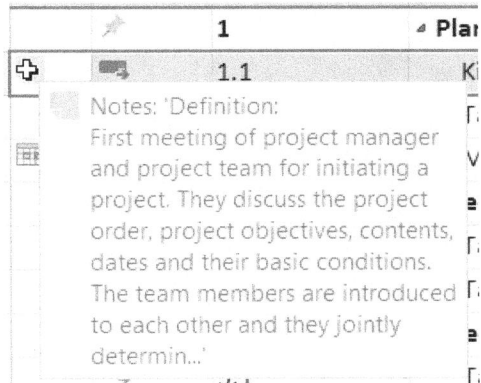

NOTES, COMMENTS:

5.7   SCHEDULING/CREATING MILESTONES

A milestone is a deadline/date/due date of project planning, a special point in time by which an important intermediate result will be reached. A milestone can only be exceeded when the previously formulated milestone requirements have actually been met. The key milestones are e.g. the transitions from one project phase to the next.

To display a milestone in Microsoft Project, a **duration of 0** must be entered. Alternatively, a task can be defined as a milestone through task information. This is represented by a square standing on its tip. By default, the date is displayed on the right.

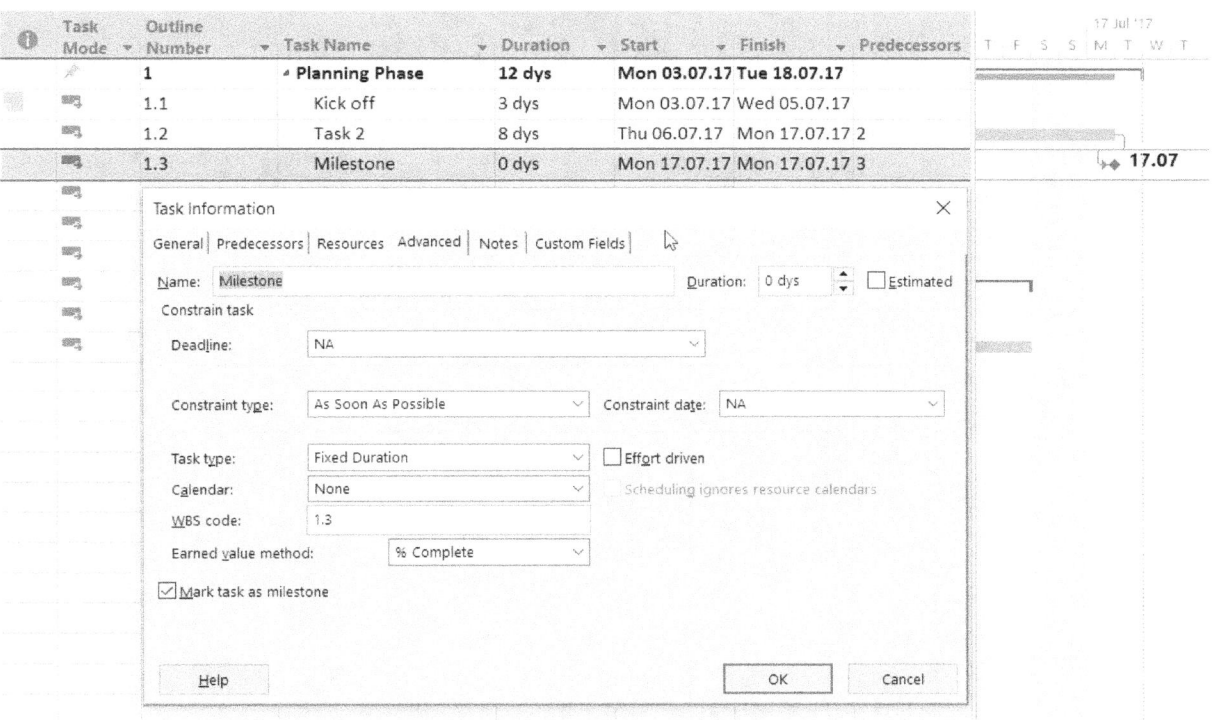

The functions "Filter" and "Group by" can be used to provide an overview of all milestones (e.g. milestone trend analysis).

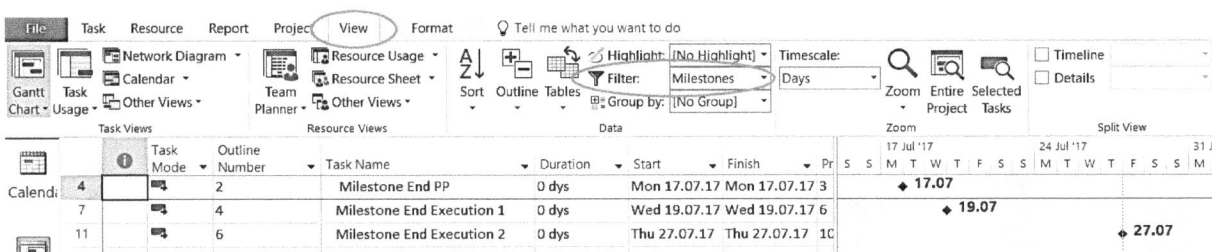

NOTES, COMMENTS:

5.8   TIMELINE

Sharing project information with users that don't have any version of Microsoft Project has been difficult and unclear for a long time as there isn't any free viewer for Microsoft Project available.

Alternatively, the project plan was transferred to a PDF format or, failing that, a screenshot was generated. Both was only suboptimal. Since version Microsoft Project 2013, the function "Timeline" is available.

The timeline, which is automatically displayed through other views, provides a precise overview of the entire schedule. Tasks can be added to the timeline, the tasks can be individually formatted and the timeline can be printed. In case the timeline isn't automatically displayed, the function can be displayed using the menu item "View". It's also possible to insert the timeline in an email and send a quick overview to all project participants, or to transfer the timeline to a PowerPoint presentation. The desired function can be selected using the drop-down menu that is shown after marking the timeline.

In this example, the gray-shaded tasks were marked and added to the timeline with the right mouse button. Alternatively, use the menu item "Task/Add to Timeline".

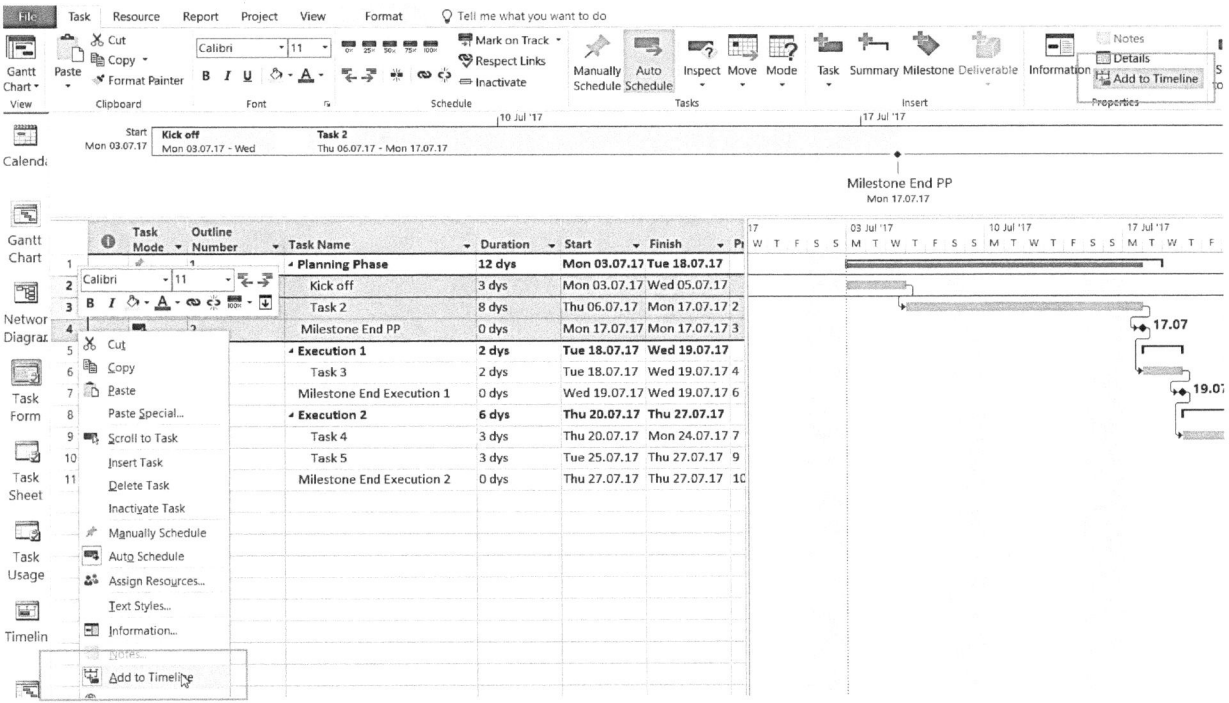

NOTES, COMMENTS:

A NEW feature in the Microsoft Project version 2016 is the possibility to display multiple timelines that show different details of the project.

In this illustration, the timeline displays some detailed tasks and the second timeline shows the summary tasks of the entire project.

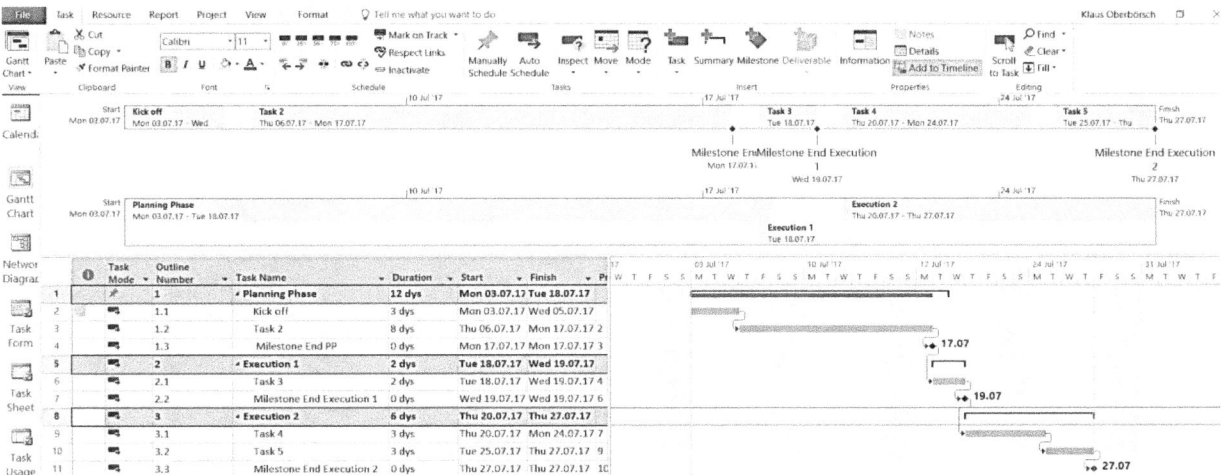

After the first timeline has been defined, click in the view area of the timelines with the right mouse button, and the following window is displayed:

An additional timeline will be generated with "Timeline Bar" where re-selected tasks are shown.

The same process is used to copy the timeline, e.g. for email (Outlook), as PowerPoint element or as a graphic. Alternatively, the command is also available in the menu bar "Format/Timeline Tools".

NOTES, COMMENTS:

5.9   CRITICAL PATH

The critical path will be defined in project management as follows:

In a project plan, the critical path is the sequence of tasks and milestones that define the minimum duration of a project. The tasks on the critical path depend on each other and have no time slacks. A project can also have more than one critical path.

Why is the critical path critical for a project?

A critical path includes all tasks that must not be delayed. In case the tasks on the critical path take longer than scheduled, the duration of the project will automatically extend accordingly. The total slack time of all tasks on the critical path is zero. If the project team e.g. needs one day longer than originally scheduled for a critical task, the entire project will automatically take one day longer.

Usually, the critical path is calculated in the form of a forward or backward pass in a network. In doing so, the free slack and the total slack are calculated. In Microsoft Project, the calculation runs in the background.

It can be presented as a network diagram, selection via "Task/View", or right mouse button at the very left of the window in the view "Gantt Chart".

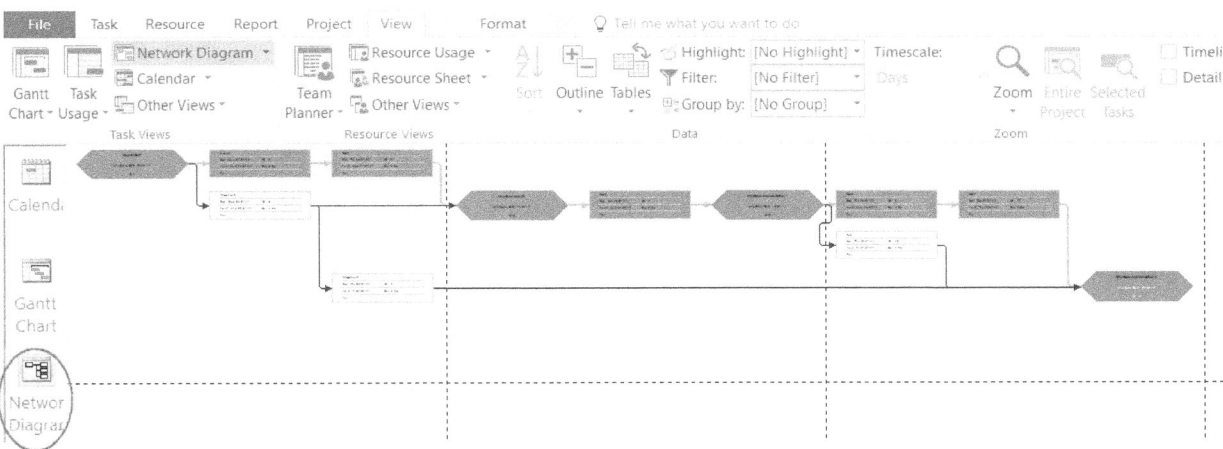

The network diagram can be enlarged using the zoom slider. The critical path is highlighted in red in the network node (in this screenshot on a gray background). All information of the network can be customized under "Format".

NOTES, COMMENTS:

Alternatively, the critical path can also be shown in the "Gantt Chart". Click the checkbox "Critical Tasks" in the menu item "Format". Here, you can also adapt formatting, e.g. to make a black and white printout.

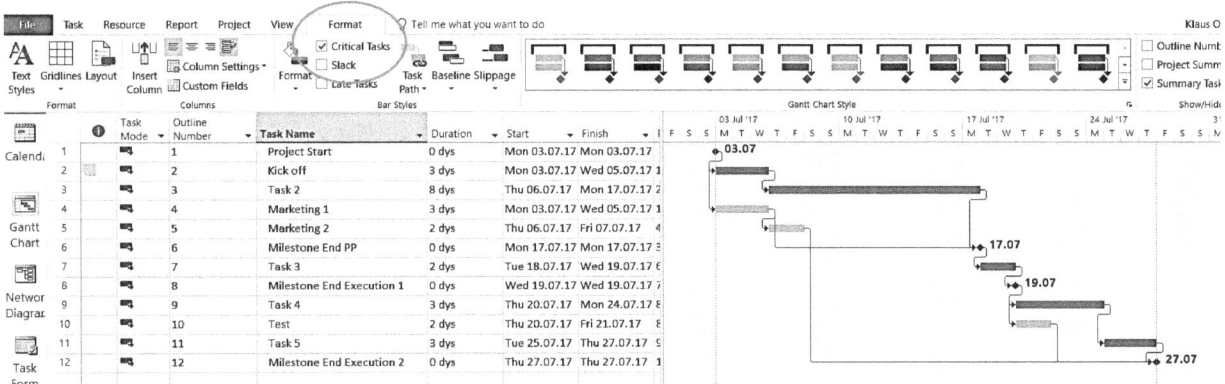

Detailed information on the tasks on the critical path such as earliest start and finish dates as well as latest start and finish dates are displayed in a special table in Microsoft Project. Change the table through "View/Tables" or select "Tables/Schedule" in the left upper corner of the table with the right mouse button.

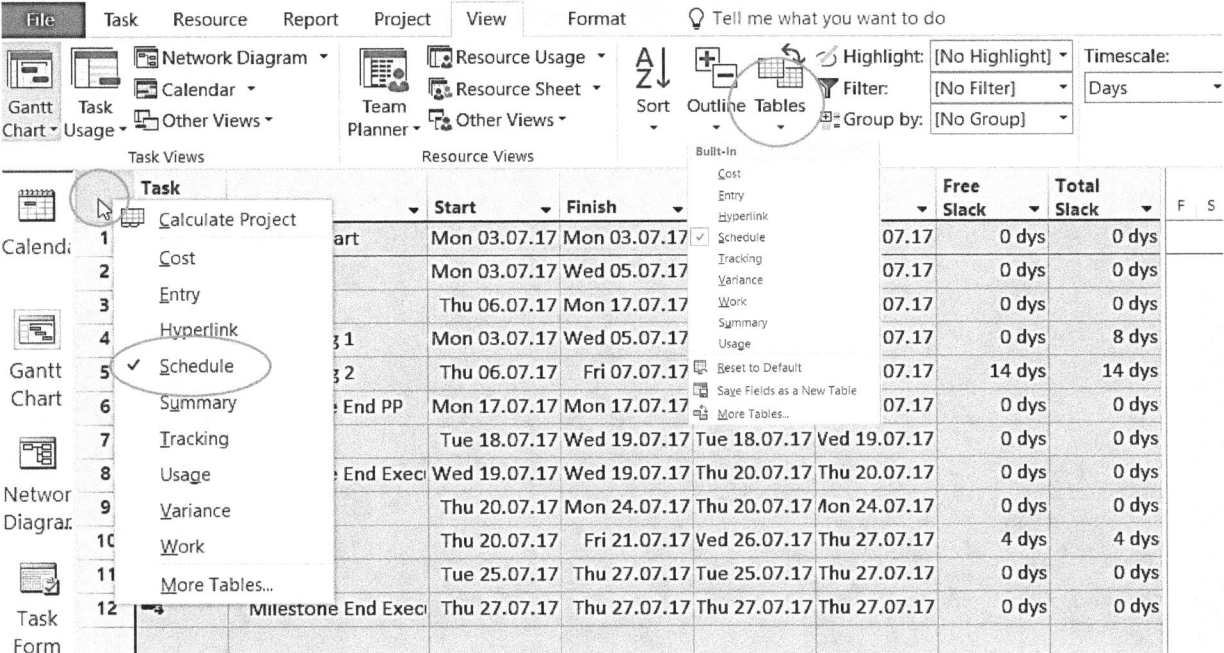

NOTES, COMMENTS:

5.9.1 SLACK TIMES

All information on the critical path are now available in the table "Schedule". Apart from the earliest start and finish dates as well as the latest start and finish dates of each task, the free and the total slack are shown as well.

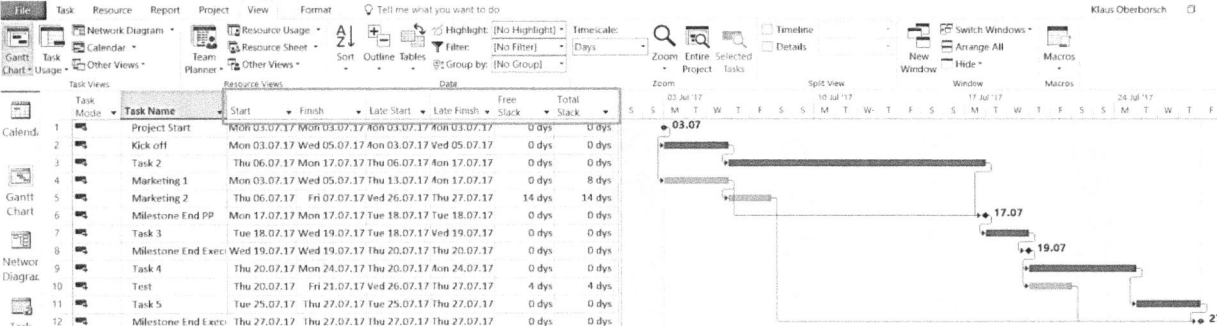

Slack is defined as follows:

The slack time is a temporal leeway for executing a task, the so-called time reserve. This leeway can be used by moving the task and/or by extending (stretching) the duration of the task.

Based on several requirements, different types of slacks time can then be determined in the network diagram.

The total slack of a task is the timespan in which a task can be moved compared to its earliest start (or duration) without jeopardizing the project finish. A task is critical when the total slack is zero.

The free slack is the time that doesn't jeopardize the earliest possible start or finish of the successor. It can only occur if at least two completed tasks meet the same successor.

NOTES, COMMENTS:

To recognize the slack in the bar chart also visually, the existing slack is displayed with a black line by checking the checkbox "Slack".

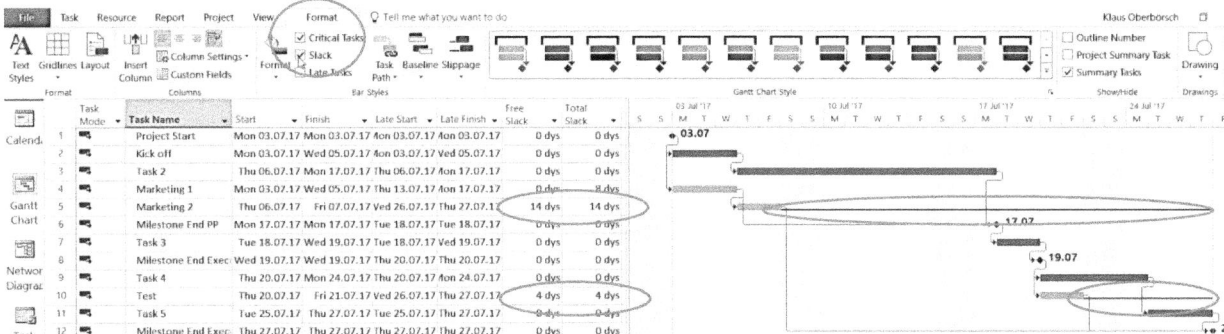

The bar can be customized using the menu item "Format" – "Bar Styles", or additional markings such as "Free Slack" on the left of the bar can be displayed.

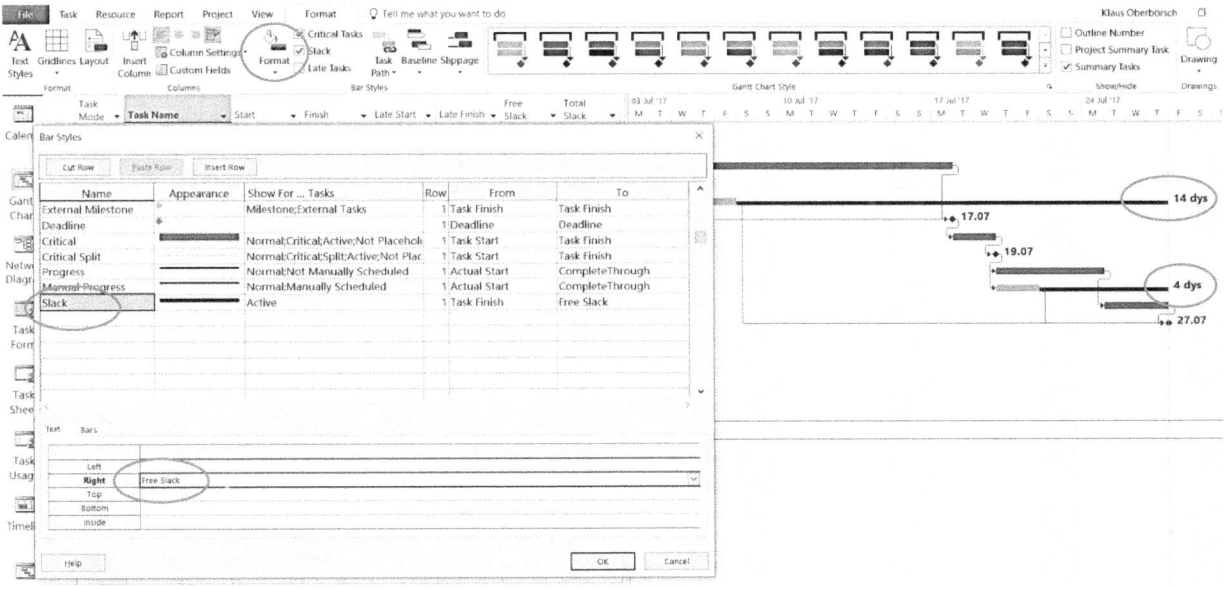

NOTES, COMMENTS:

6 TABLES

The left area in the view "Gantt Chart" is referred to as table area. Different data fields, in which Microsoft Project saves data, are shown in a table. Data fields are grouped by themes in built-in/default tables.

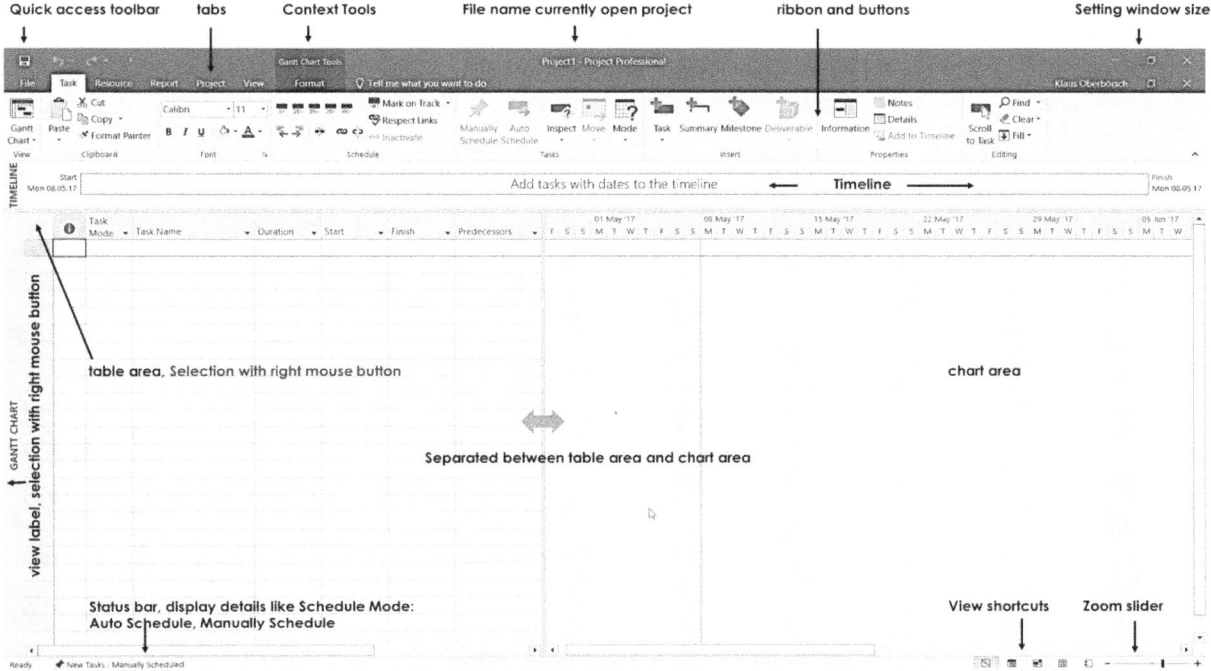

6.1 STANDARD TABLES

The following standard tables are available:

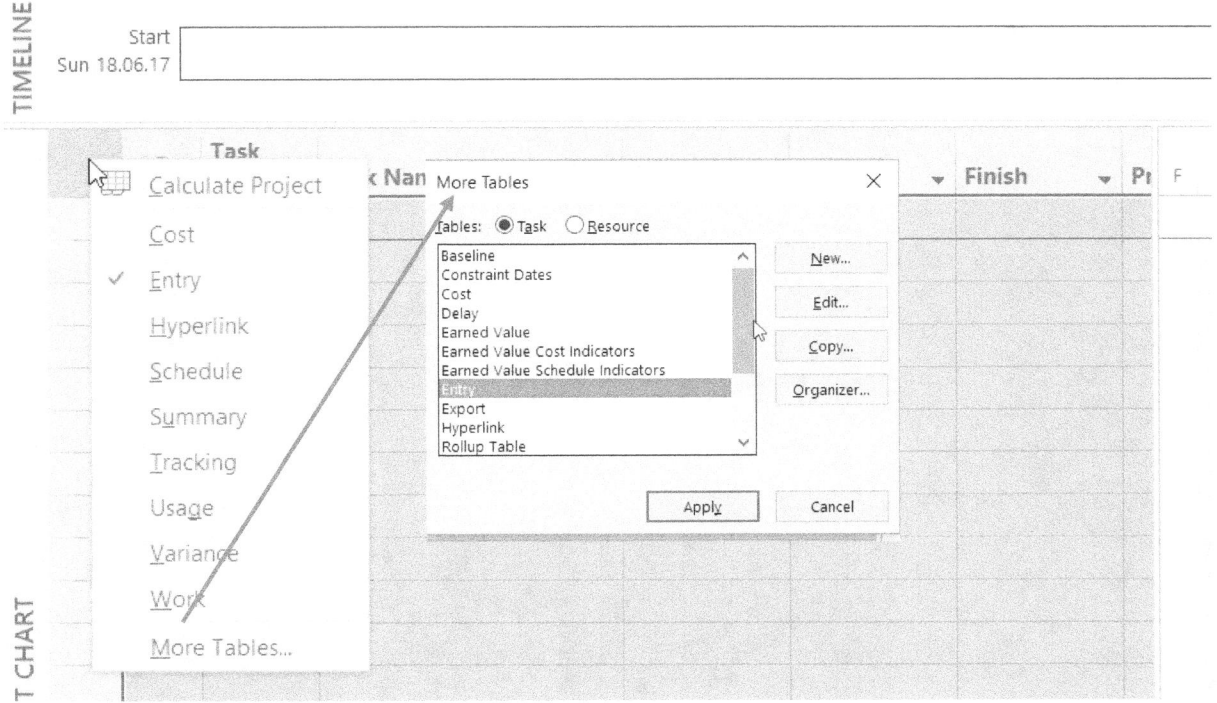

NOTES, COMMENTS:

6.2 MORE TABLES

Under "More tables" you can find a selection of tables with special information that can also be modified (more columns, custom fields …), later saved under individual names and displayed in the menu.

The definition of special tables is particularly helpful and provides an overview for customized adaptions.

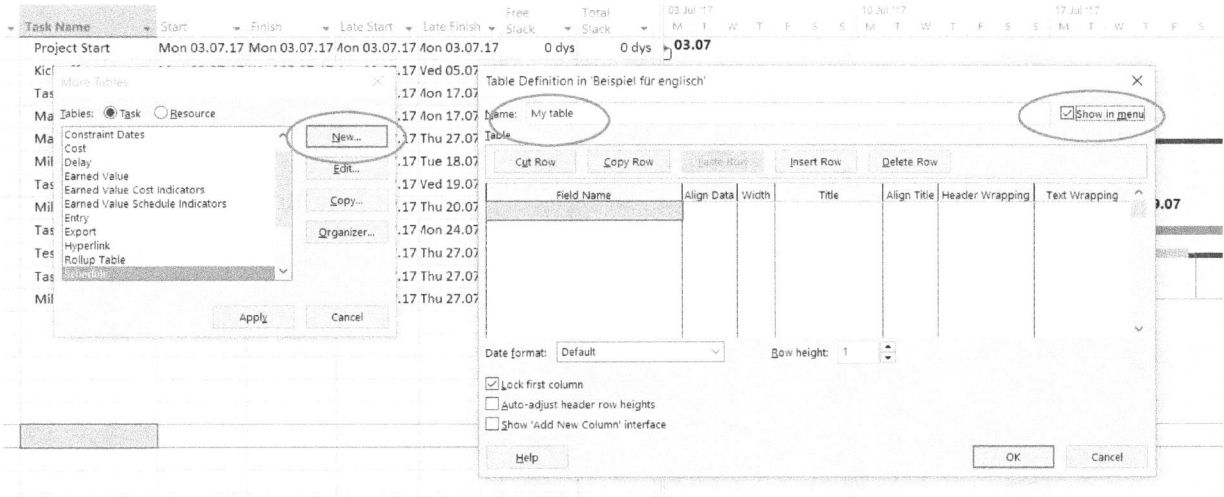

NOTES, COMMENTS:

7 SCHEDULING RESOURCES

Different resource types are used for completing or processing tasks: Persons and/or equipment. As of version Microsoft Project 2007, so-called cost resources can also be assigned to tasks in order to directly assign costs such as e.g. travel expenses to a task (see below section 8.1)

After scheduling assignments/tasks, the required resources are assigned based on tasks.

The resource assignment (assignment of resources to tasks) enables you to:

- Show which employees should participate in specific tasks at a certain time
- Calculate the effort and expenses for the resource usage
- Identify overallocated resources
- Display free capacities of resources to assign new tasks

7.1 SCHEDULING RESOURCE USAGE

There are various ways to capture and manage resources:

- Project-based
- Across different projects (creating a resource pool)
- Company-wide (creating an enterprise resource pool – using Microsoft Project Server and Microsoft Project Professional)

Which method is most suitable for you, depends on how your company uses Microsoft Project.

In the first variation (project-based), you insert all necessary information in a project plan (i.e. in a file). This is for example the case when a project manager works with a project plan all by himself. In this case, you first collect all resources in a resource table including all information and assign them to the tasks.

In the second variation (creation of a resource pool within the scope of a multi-project management scheduling), not only one project is managed but several projects at the same time. The resource assignment and utilization can be identified across different projects. Usually, the projects are even managed by different project managers. In those cases, the resources are inserted in a project file. Each project is linked to a resource file. This way, you can quickly recognize if the desired resources still have free capacities.

NOTES, COMMENTS:

In the third variant (creation of an enterprise resource pool by Microsoft Project Server), all enterprise resources, either employees or material, are saved in a designated pool that can also be protected by an assignment of rights in terms of access and use of resource data.

The resource can be selected and assigned separately for each project. Apart from the dedicated assignment of rights, the enhanced ease-of-use compared to the second variation (resource pool) is another great advantage. The use of the enterprise resource pool requires Microsoft Project Professional and Microsoft Project Server with web applications for users and resources.

Below, we will take a look at the first variant (project-based).

7.2   INSERTING AND MANAGING PROJECT BASED RESOURCES

The view "Resource Sheet" serves to record resources. You call them up by clicking with the right mouse button at the very left of the window. All possible views will be displayed.

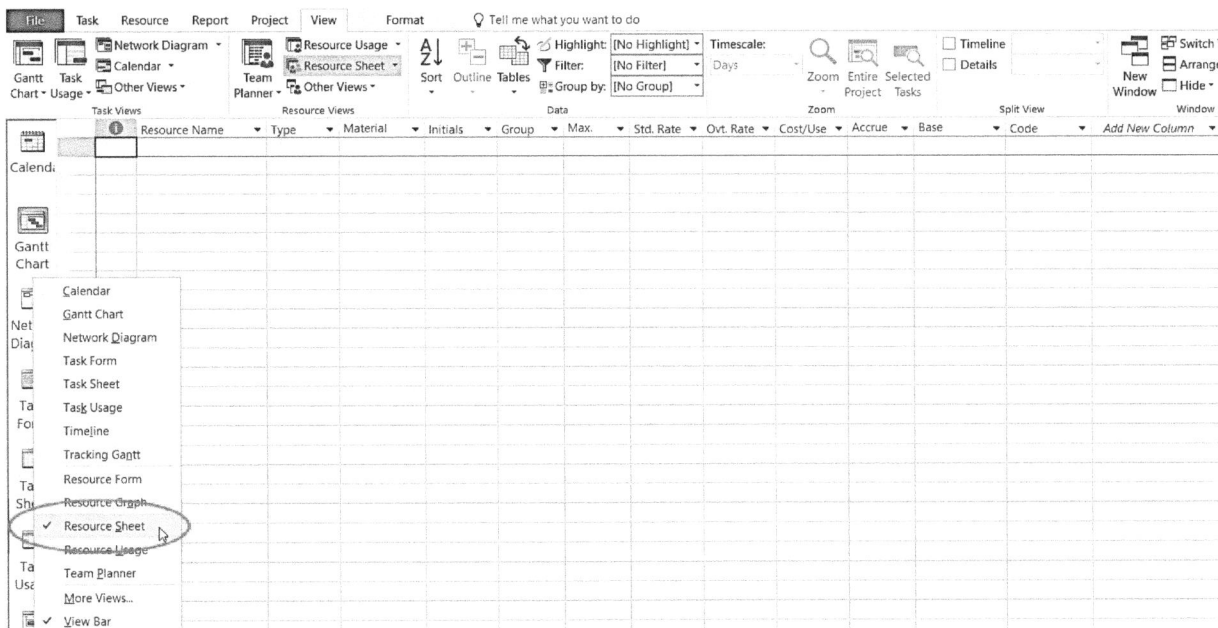

Alternatively, select them via the menu item "View/Resource Sheet".

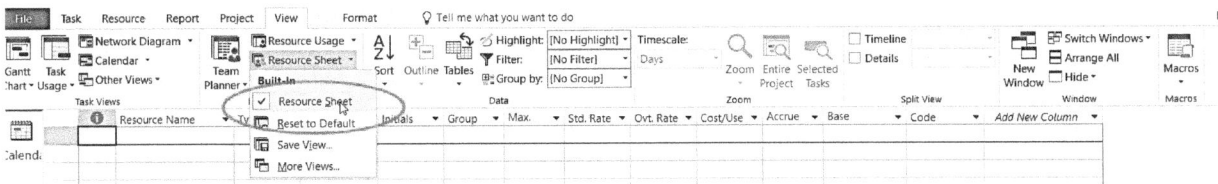

You have to capture the resources in the table which appears. Alternatively, you can use copy and paste from another application or an Excel import.

NOTES, COMMENTS:

The resource name will be entered in the corresponding column, structure of the first/last name according to company regulations. The resource names will be used in different views, meaning that plain text or explicit descriptions are recommended.

Microsoft Project can distinguish between 3 different resource types (important for evaluations, groupings, filter...). Traditional work resources in the form of persons should be entered as work.

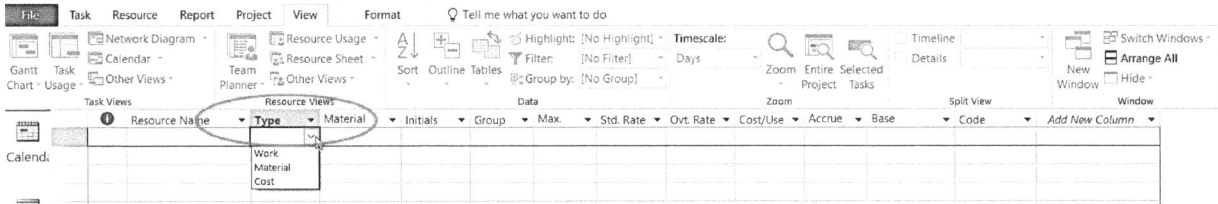

The selection of material resources is useful when it comes to resources that aren't necessarily charged by hours but by units – such as e.g. material like sand, one-time rents for rooms or machines, staff that is paid as a lump sum.

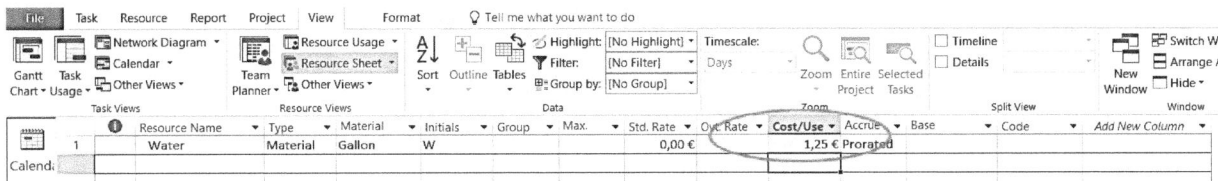

Costs per task can be defined by cost resources, such as e.g. travel expenses, hospitality costs, i.e. costs that can vary by task.

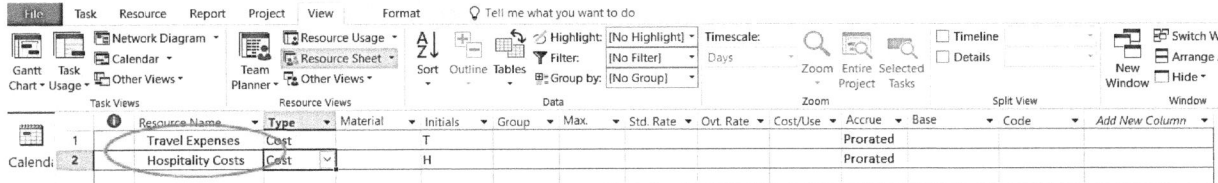

The possibilities to evaluate and consolidate will be described under cost management (see section 8.1 below)

NOTES, COMMENTS:

The column "**Material Label**" serves to classify material. Here, you can enter standard values like cbm, liter etc. that are shown under the respective material resource.

In the column "**Initials**", initials can be used that later serve as an alternative for the whole resource name (bar labels or the like).In case you don't enter any initials, Microsoft Project automatically uses the first letter of a name.

In the column "**Group**", resources can be assigned to different resource groups (own information is necessary) such as internal/external, departments, project teams etc. To achieve a consistent data base, it's, however, recommended to insert data through a custom field with options, see chapter 11.

The column "**Max. Units**" allows you to document the maximum availability of the resource. In general, the calculation is based on 100% (= full working time). If the person is only assigned for half of this project or when it comes to a part-timer, this information is stored by entering 50% (in decimal mode 0.5, can be changed under "Options")

Insert the **standard rate** as well as the **overtime rate** in case you want to plan costs for a project. Any amount that you enter is automatically interpreted as an hourly rate. If you would like to insert a daily rate, you have to use a daily rate (example: €600/day).

In the column "**Costs/Use**", you can enter one-time costs for work resources (flat-rate travel expenses or other one-time costs that are charged apart from hours worked). You can enter the costs for the cost type "Material" in the same field that are then charged as one-time costs per usage, independent of the performed units.

If needed, you record the field "**Code**" which also can be used to insert cost centers.

Additional information on resources can be entered through "**Resource Information**" You call it up by double-clicking on the resource row on or via the button in the toolbar.

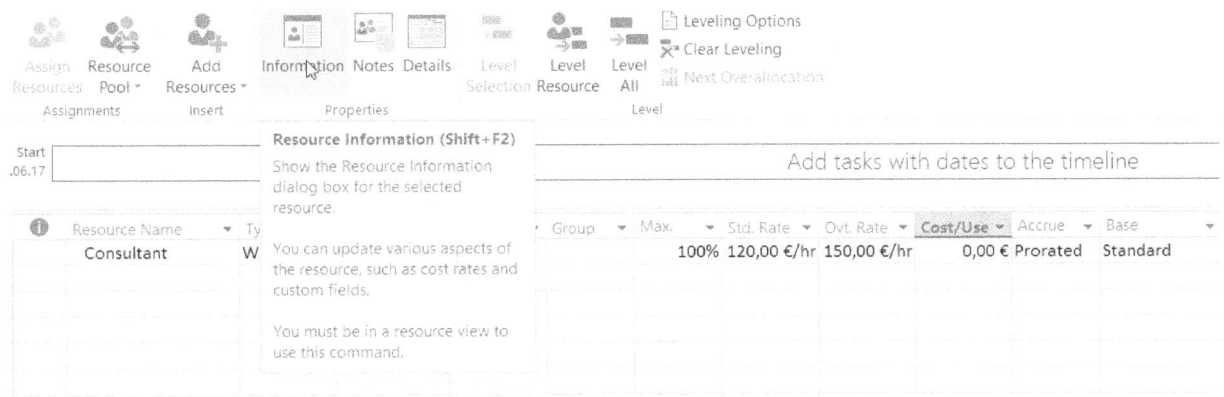

NOTES, COMMENTS:

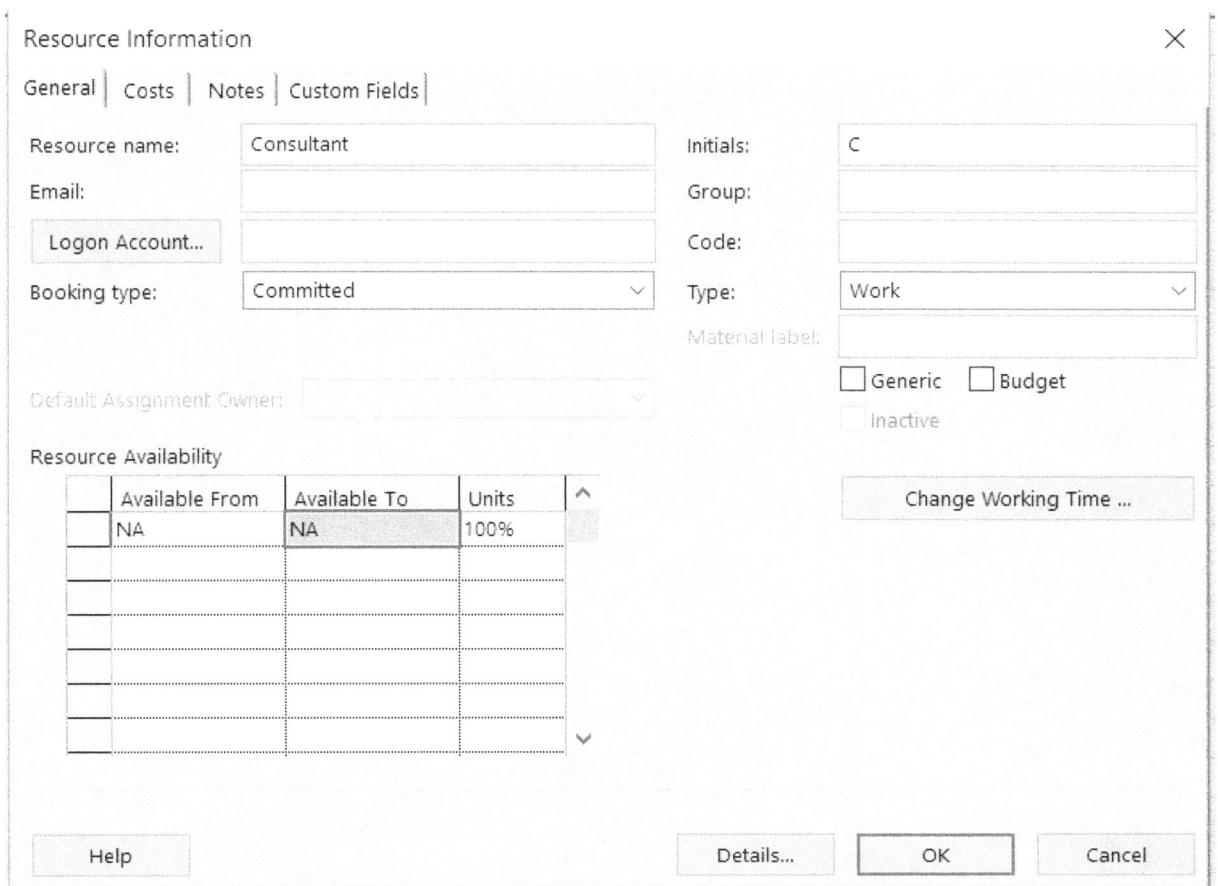

Other information that you can add in the window "Resource Information" include the field "Email". As soon as it has been filled out, project information can be exchanged between project participants using the workgroup function of Microsoft Project (a configured email server such as Exchange or the like is required). The "Resource Availability" or the constraints due to vacation etc. can be entered in this window as well. The individual working time per resource is set using "Change Working Time ..." as well as the feature "Budget" resource.

NOTES, COMMENTS:

7.2.1 ASSIGNING RESOURCES TO TASKS

Allocating a resource to a task specifies which resource performs a certain task or which material is required. You can also determine how much effort (work) a person contributes to this task or at what utilization percentage a person participates, if needed.

Microsoft Project provides different ways of assigning resources. The optimal way to assign a resource takes place by accessing the menu item "Resource/Assign Resources". A window with the available resources opens, all resources integrated into the resource table are shown here.

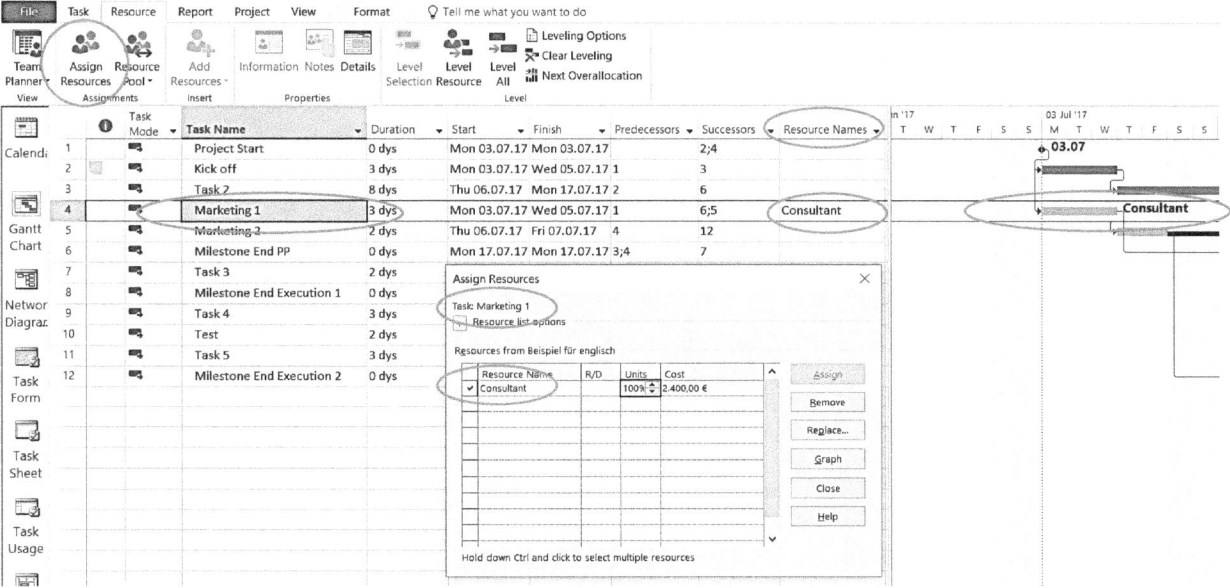

The task that should be assigned to a resource is marked. The window "Assign Resources" allows you to select the desired resource by inserting the required percentage of the unit under "Units" via the drop-down menu. You can assign the resource to the selected task by mouse click in another field or using the button "Assign". Indicated by the resource names behind each bar in the Gantt chart (values other than 100% are shown, no percentage value = % assignment of 100%) The resources assigned to a task are always on top of the shortlist. Furthermore, the resource assignment is displayed in the corresponding column in "Table: Entry".

The costs incurred by the resource assignment are charged in column "Cost" ("Table: Cost"), also the created work in the "Table: Work".

NOTES, COMMENTS:

7.2.2 Display Resource Usage

Detailed information on the resource assignment can be found in the view "Resource Usage". It focusses on the resource perspective. Each scheduled resource is shown in detail in the respective tasks. Here, you can carry out a detailed scheduling per day to offset small overloads, if necessary.

If you check the checkbox "Details" and select "Gantt Chart", you receive an individual Gantt chart depending on the marked resource.

The detailed view "Task Mask" enables you to change the working hours and the general assignment to a task.

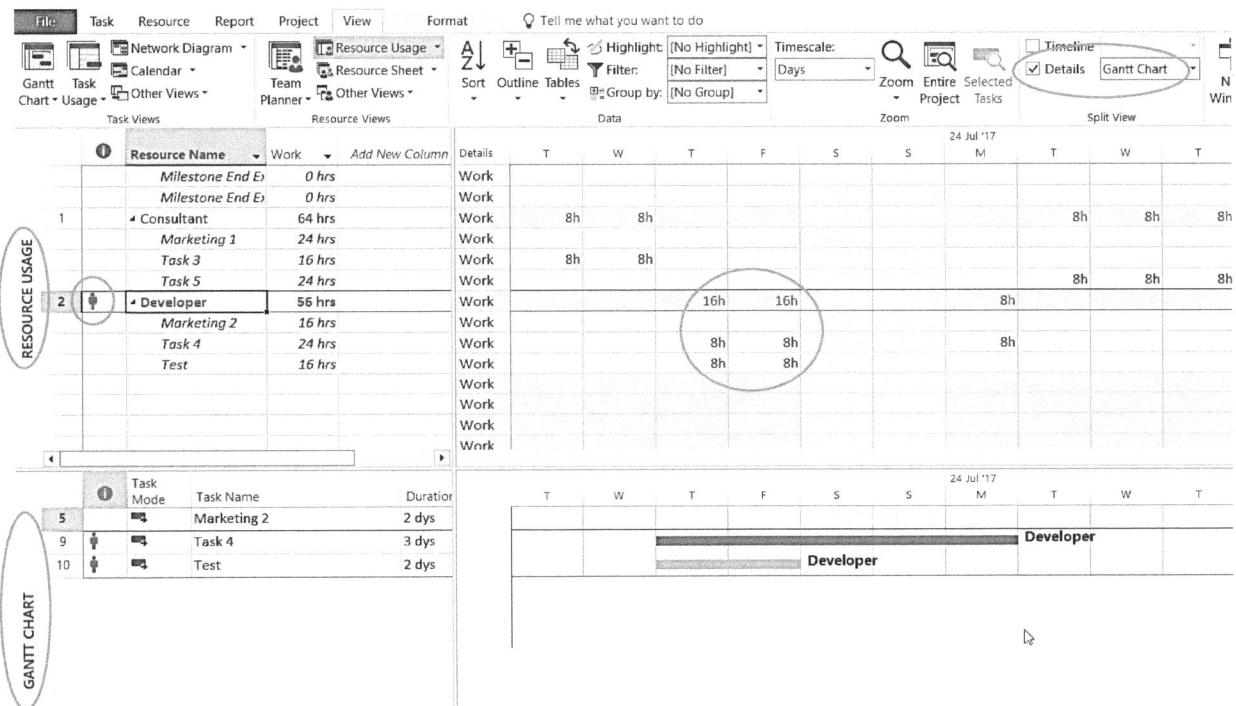

Microsoft Project always allows the assignment of resources to new tasks, even if the resources are already overallocated or overallocations have be caused by a new assignment. However, in case of an excess capacity of resources, Microsoft Project shows where the overallocation occurs so that you can change the assignments accordingly. Alternatively, you can also use the resource leveling function. Resource leveling causes a rescheduling of tasks in consideration of additional conditions (e.g. priorities) so that resources are relieved.

NOTES, COMMENTS:

When allocating resources in Microsoft Project, a differentiation must be made between the first assignment of resources to a task and later assignments. In case no resource has been assigned to a task, Microsoft Project always calculates the total effort during the first resource assignment according to the following formula:

Work = duration * units * 8 hours/day

In case of an additional resource assignment (or withdrawal), the changes in duration and work depend on the defined task modes as well as the effort driven (further explanations can be found in the next subsection "Task Modes")

Task modes as well as effort tracking reflect different situational functionings. Examples: When another bricklayer is added to a productive task such as e.g. brick-laying, the work effort generally remains the same but the duration for the task is reduced. When another software tester is added to a five-day software test, the quality usually improves as a qualitatively better result can be reached with a higher work effort. Consequently, the duration remains the same, the total work effort increases.

7.2.3   TASK MODES AND EFFORT TRACKING

Task modes and effort tracking define how the other parameters change in case one of the parameters such as duration, work or resource unit changes. The changes are situational. Therefore, the task mode of a task has to be changed several times during a project.

There are three task modes in Microsoft Project:

- Fixed units
- Fixed duration
- Fixed work

In the default setting of Microsoft Project, new tasks are set to the task mode "Fixed Units/Effort driven". The current presetting can be recognized and adapted under "File/Options/Schedule" depending on the requirements in global.mpt.

NOTES, COMMENTS:

Fixed units, effort driven

Resources have already been assigned to a task. Now, another resource is assigned. The percentage of the assignment to the task remains unchanged. The duration of the task is reduced accordingly.

Example:

To create a user manual, two employees need 9 days at full capacity (100%). Based on the formula for work this results in:

Duration of 9 days x 200% units x 8 hrs/day = 144 hours of work.

As a result, each employee works 72 hours.

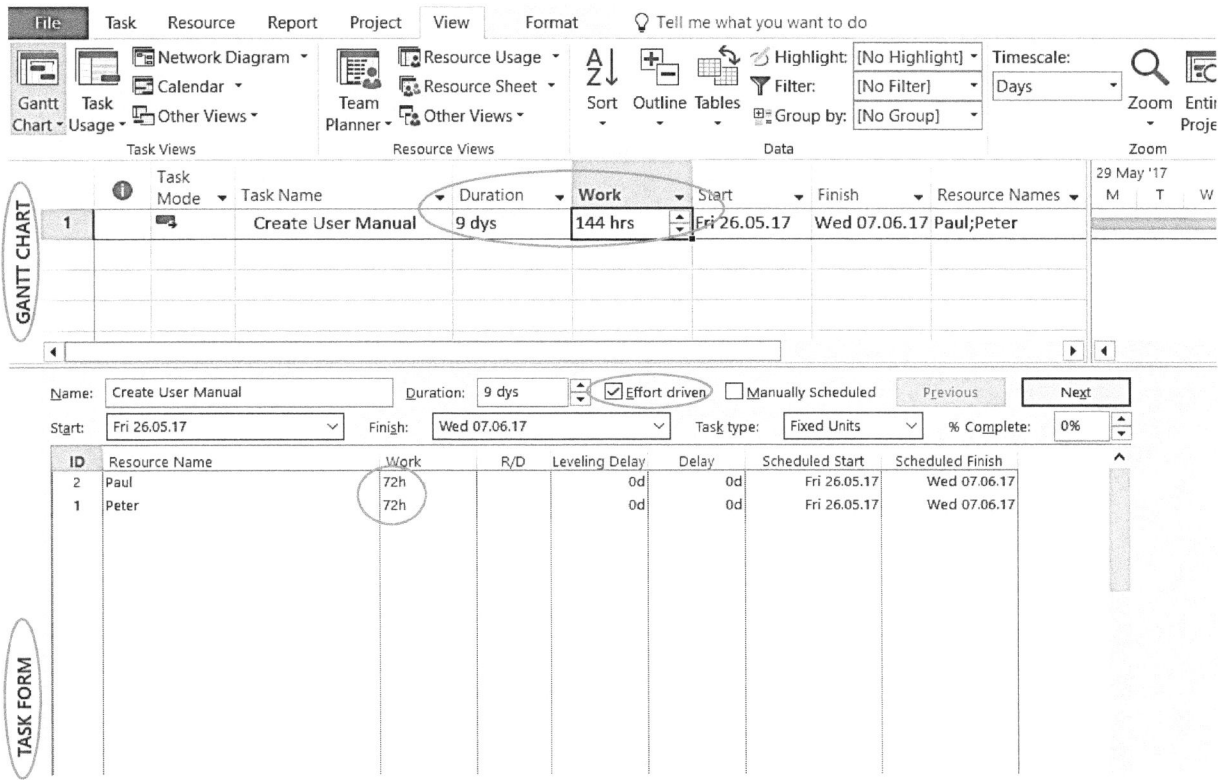

NOTES, COMMENTS:

If an additional employee is added, these hours of work are distributed proportionally to the three of them. Each employee invests 100% of his manpower but only works 48 hours. The work is completed in 6 days.

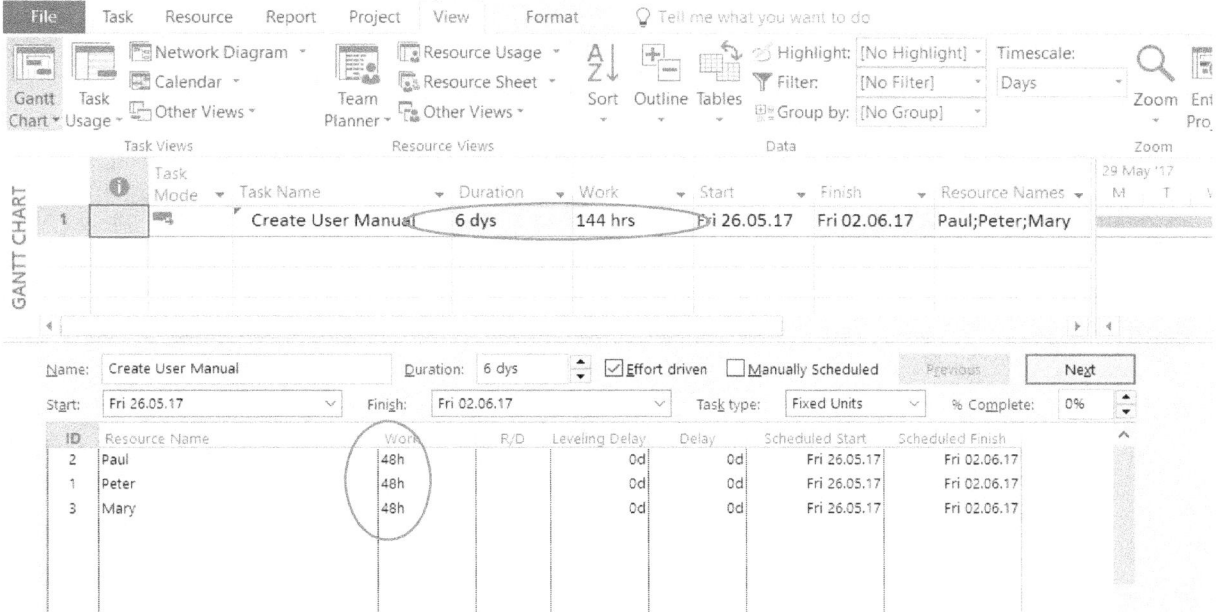

NOTES, COMMENTS:

Fixed units, non-effort driven

There can be situations in a project that belong to the category "Fixed Units" but are non-effort driven.

Example:

You have planned an expert panel regarding a specific problem. When assigning an additional expert, the problem is discussed more thoroughly and other aspects can be considered. However, adding another expert doesn't result in a shorter duration but of course in an increased effort.

Deactivating or activating the effort tracking "Effort driven" can temporarily be necessary, as described in the following scenario.

If in the above example about employees creating a manual, a coordinator is assigned to those three employees with a quarter of the full working time, he surely executes some work tasks but doesn't directly support the work of the other two employees. Therefore, prior to the assignment of the coordinator, the function

"Effort driven" should be deactivated. Microsoft Project now adds the workload of the coordinator so that the overall workload increases by 12 hours. The duration of the task remains unchanged at 6 days.

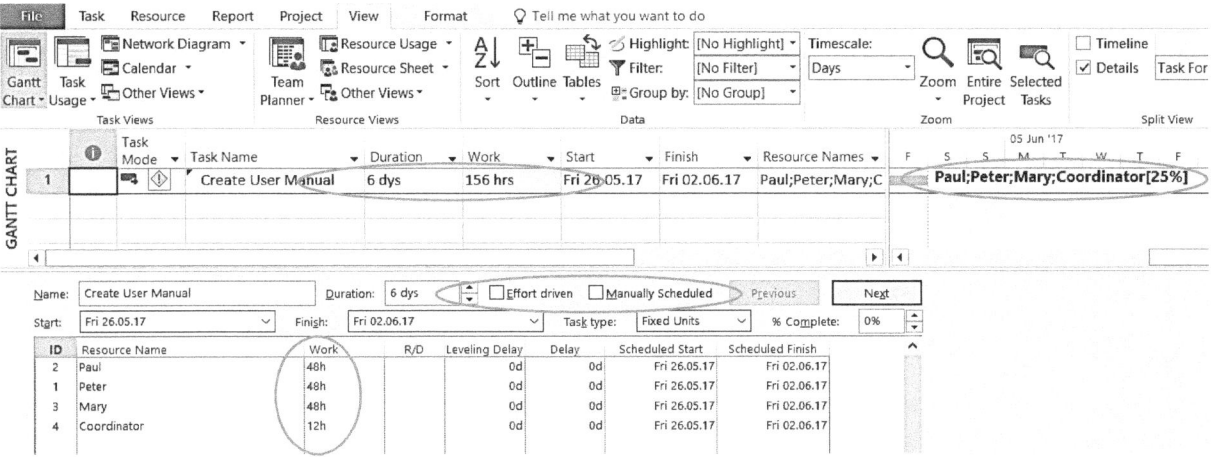

NOTES, COMMENTS:

Fixed duration, effort driven

When setting a fixed duration, the duration of the task remains unchanged. You can change the units of the task or the effort (work). The other value will then be recalculated.

Example:

Three employees have to test a new software for two days according to specified test scenarios. The duration is defined as a fixed period of time.

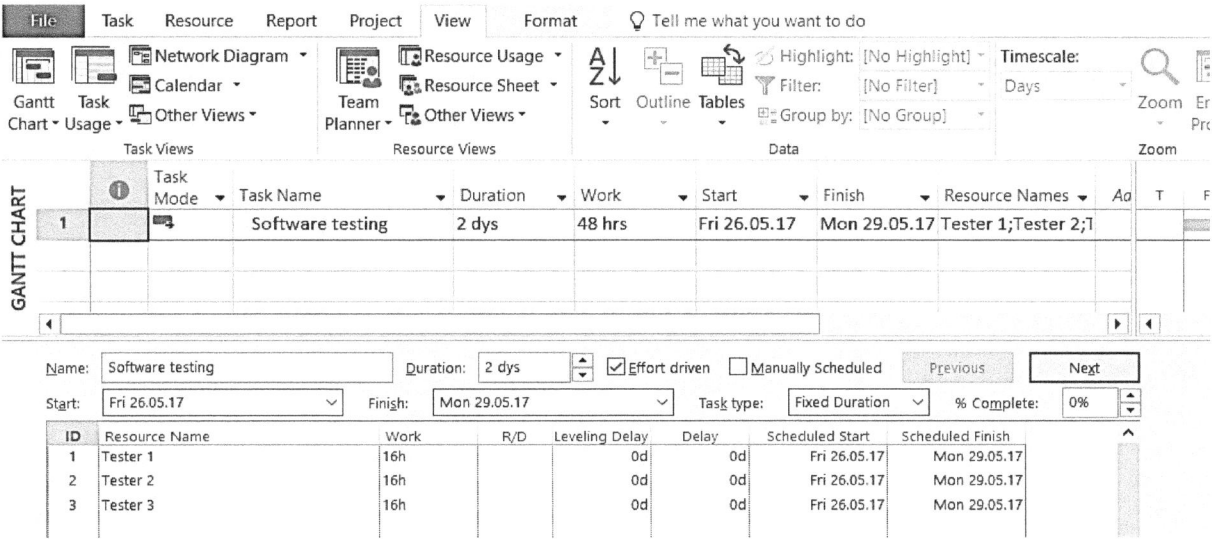

If another employee will be added as a tester, the test still takes two days due to the defined duration. However, the tester isn't working at full capacity anymore as the test scenarios are completed earlier.

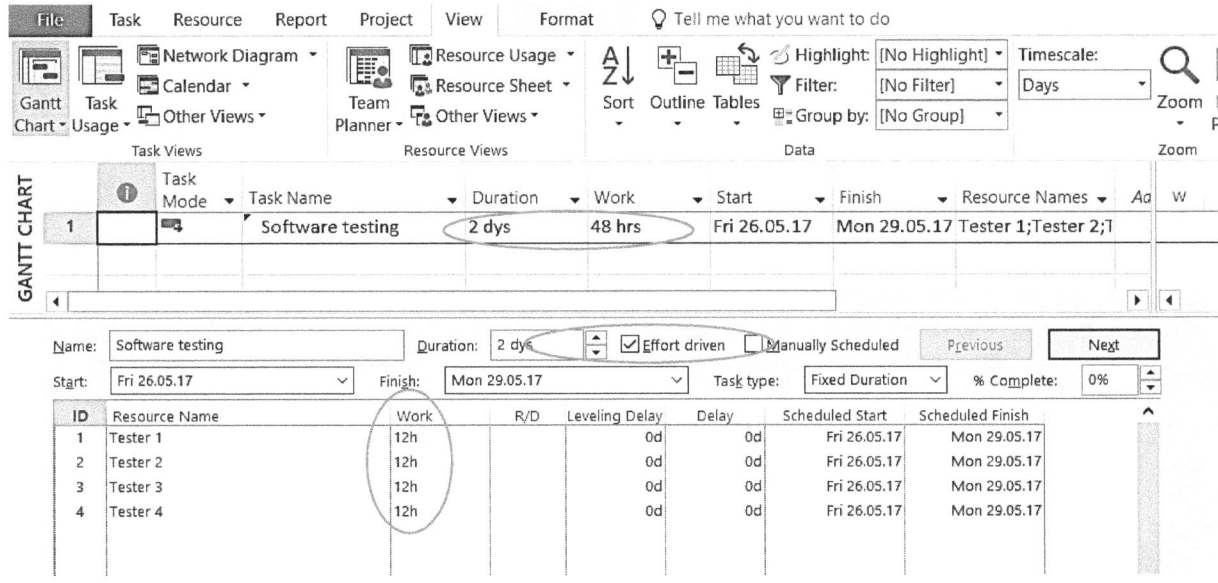

NOTES, COMMENTS:

If, in case of the software testers, the tests weren't based on test scenarios but rather on free testing, adding another tester would improve the quality of the results. The work effort would increase since the duration of the test would still contain two full working days for all testers.

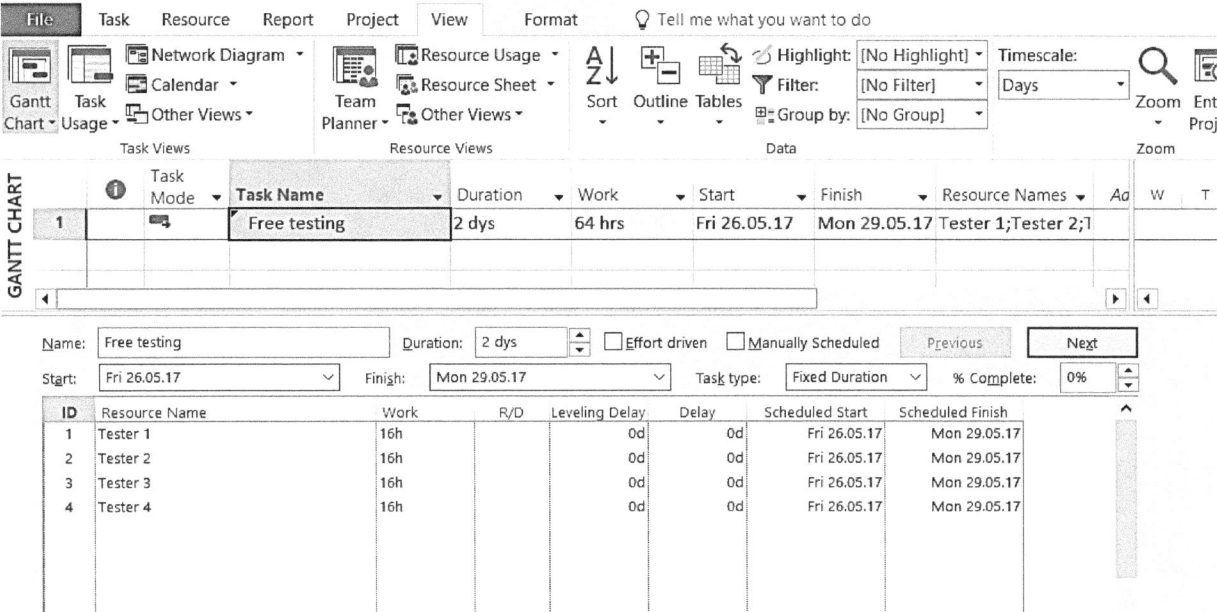

Deactivating "Effort driven" implies that the work effort increases when assigning another resource.

NOTES, COMMENTS:

Fixed work, effort driven

When setting fixed work, the duration of a task remains unchanged. You can change the units of a task or the value for the duration. The other values will then be recalculated. "Effort driven" cannot be deactivated for this task mode.

Example: 10 employees need 20 days to clinker a house. If you add other employees, the duration is reduced accordingly.

How task modes can affect your schedule:

Microsoft Project uses a planning formula that puts the three values work, duration and assignment units into relation with each other: **Work = duration x units**

Below, you will find a table including an overview of all variable elements of the scheduling formula:

In a	If you revise units	If you revise duration	If you revise work
Fixed units task	Duration is recalculated.	Work is recalculated.	Duration is recalculated.
Fixed work task	Duration is recalculated.	Units are recalculated.	Duration is recalculated.
Fixed duration task	Work is recalculated.	Work is recalculated.	Units are recalculated.

NOTES, COMMENTS:

7.3 TEAM PLANNER

Another, very clear option to assignment resources is included in the function "Team Planner". The function is located under menu item "Resource" on the left.

In this divided view, the resources and the tasks assigned to them are displayed in the upper area. The lower area shows the tasks that haven't been assigned yet in the form of a Gantt chart.

You can now drag the task to the relevant resource with the left mouse button. The task will then be removed from the lower area.

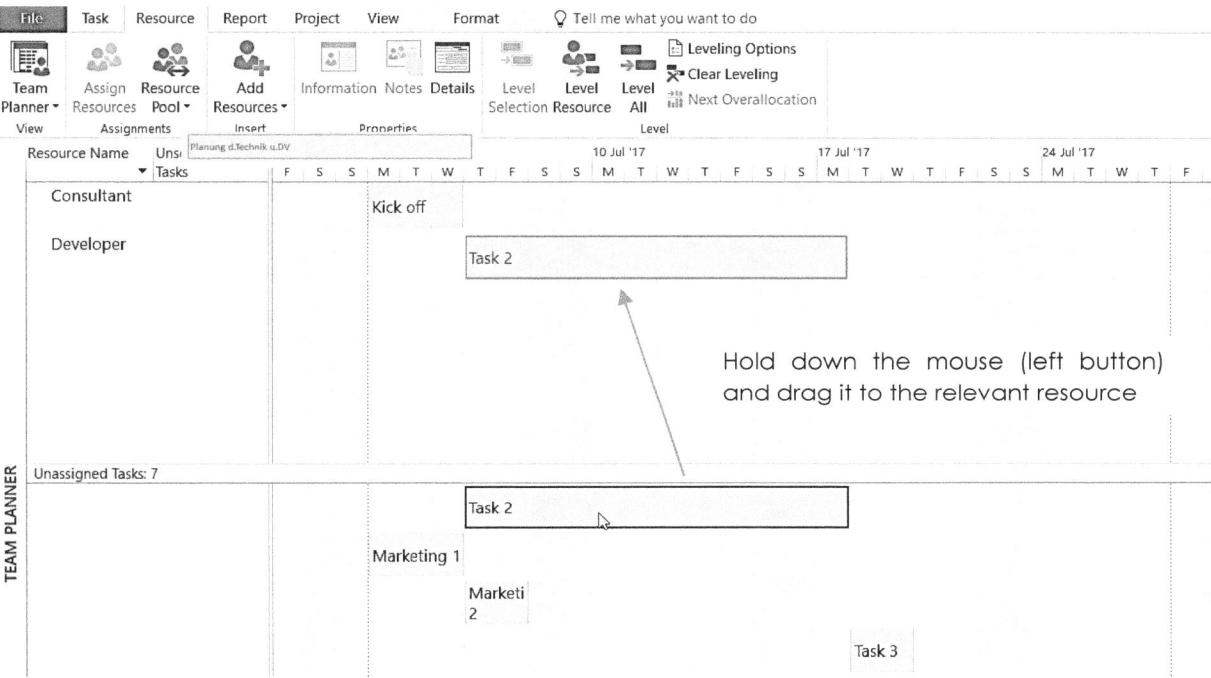

Of course, also multiple tasks of one resource can be assigned, possible overallocations are indicated by the resource name highlighted in red and by a red bracket around the overlapping tasks.

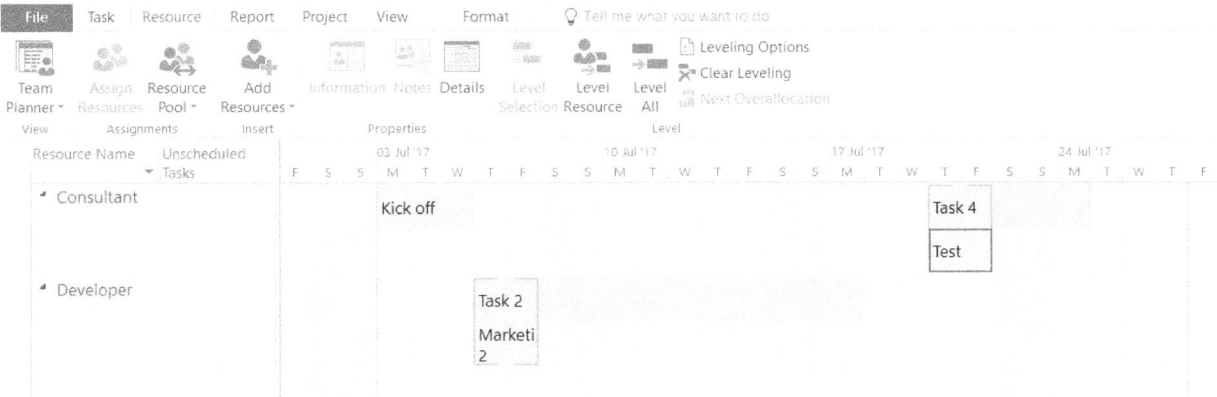

NOTES, COMMENTS:

7.4 RESOURCE LEVELING

7.4.1 AUTOMATIC RESOURCE LEVELING

Microsoft Project includes the feature "automatic capacity leveling" allowing that overallocations of resources can automatically be cancelled/re-planned. This is can be used for single resources as well as for all resources of the project.

If e.g. two tasks run in parallel and both tasks were assigned to the same resource at 100%, this resource is overallocated. As a result, the start of one of the two tasks is delayed/moved until the resource is no longer overallocated. In this example, the resource Uli Lauterbach has been assigned to two tasks at the same time (selection of DP and selection of telephone system) that are also scheduled simultaneously. The overallocation is displayed in the indicator column.

1. Select the Resource tab

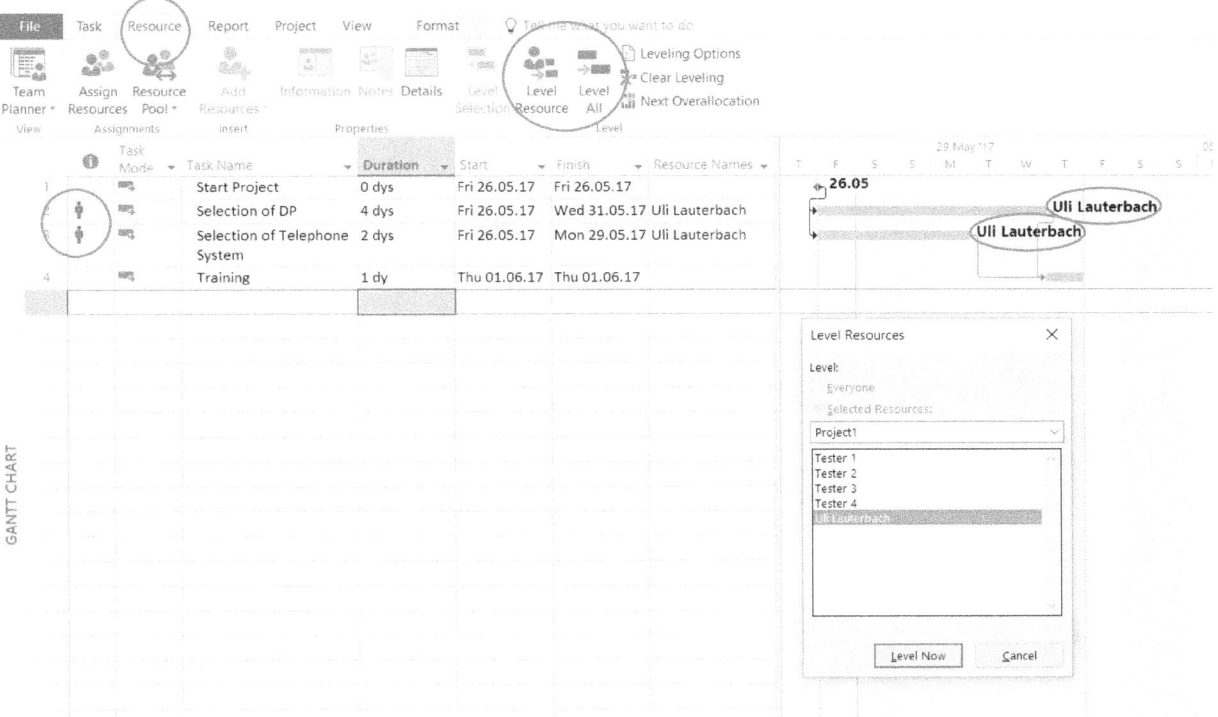

2. Select "Level Resource" or "Level All". Check/adapt possible settings in the menu item "Leveling Options" beforehand, e.g. "Level only within available slack".

3. Confirm the settings with OK and select between the options "Selected Resources" or "All". With one mouse click on "Level Now" Microsoft Project carries out a rescheduling to schedule the overallocated resource in a way that the overallocation is cancelled.

NOTES, COMMENTS:

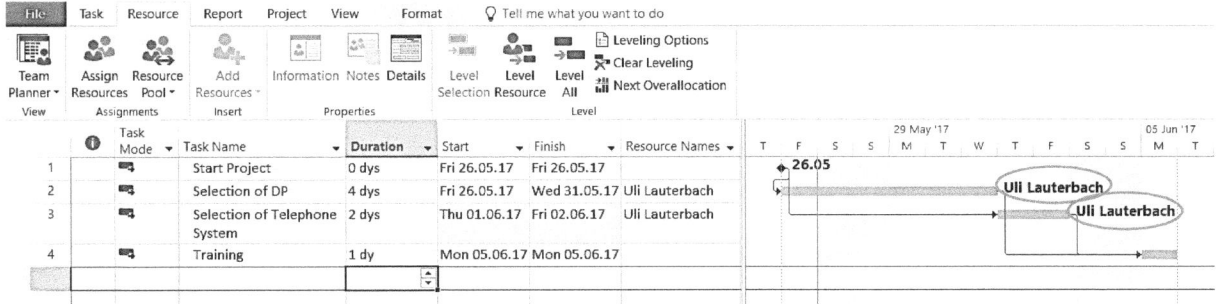

The resource Lauterbach is no longer overallocated as the second task is delayed until the first task is finished.

However, a leveling across the entire project could result in many date conflicts and associated notifications. Therefore, you should always only select a "manageable" period of time for leveling.

It's not recommended to select "Automatic Resource Leveling". Since this function is a program default, it has an effect on everything you will enter. After each entry that would involve an overallocation, a leveling would be carried out again. This means that tasks would continuously be moved.

The delay/change in the project plan due to a resource leveling can quickly be removed again using the button "Clear Leveling". This results in another overallocation of the resource(s) and another solution of the problem must be found (e.g. manual leveling of another resource).

7.4.2 MANUAL RESOURCE LEVELING

After attention has been drawn to the overallocation of a resource in the view "Resource Usage" and "Gantt Chart" through a red resource icon in the indicators column, a detailed analysis of the overallocation can also be carried out using the filter function in the Gantt chart.

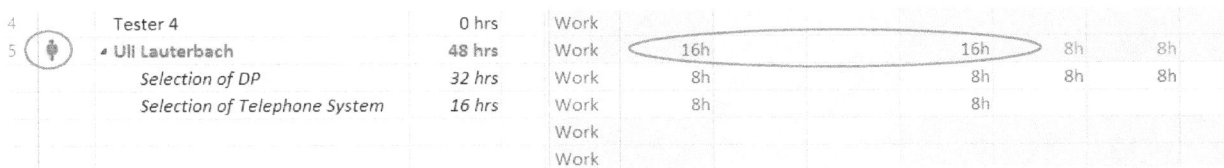

In the view "Resource Usage", the resources are displayed along with their assigned tasks. The assignment is shown by the day in the right area.

NOTES, COMMENTS:

In this case, the resource Uli Lauterbach is overallocated on 2 days (16 hours each) as he works on two tasks at the same time – once at 100% = 8 hours and in the other task also at 100% = 8 hours. Based on the default calendar, working times exceeding 8 hours are displayed as overallocations.

It's now possible to align the resource by the day by manually changing the scheduled number of hours, e.g. for each task only 4 hours per day. Note: Minimal overallocations can be modified this way. When changing the number of hours, the originally scheduled work is simply removed while the scheduled work is missing.

Alternatively, the task of another resource could be assigned to cancel the overallocation. To check a possible manual leveling in this example, we will now only take a look at the tasks related to the resource Uli Lauterbach through the view "Gantt Chart" by using the filter function.

Select the filter "Using Resource" using the filter function and then select the resource Uli Lauterbach from the drop-down menu.

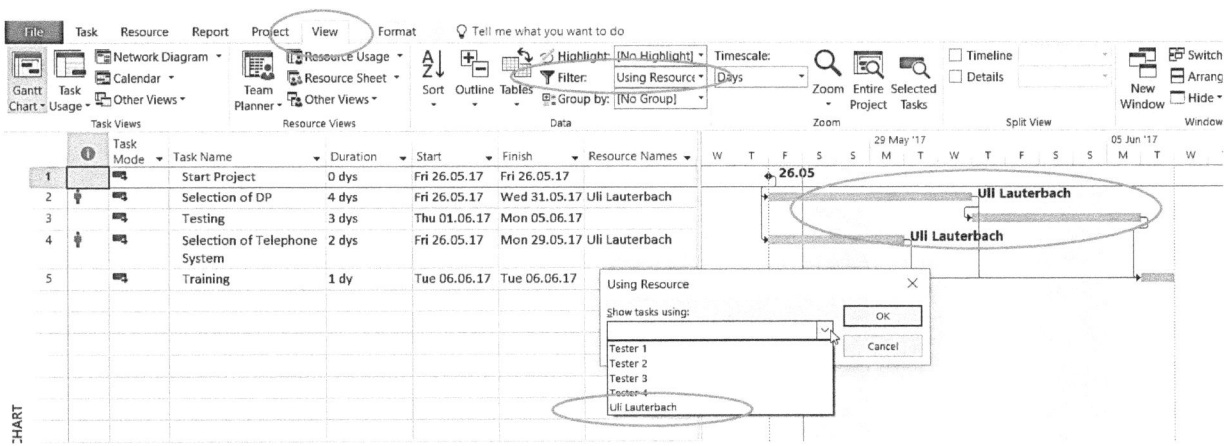

NOTES, COMMENTS:

Now, you can only see the tasks that affect Uli Lauterbach. The project planner can only take relevant measures. In this case, it's possible to move a task as it has slacks. You shift the task by holding down the left mouse button as far to the right until the overlapping of the tasks is resolved. For control purposes, a window automatically opens and displays the corresponding date values.

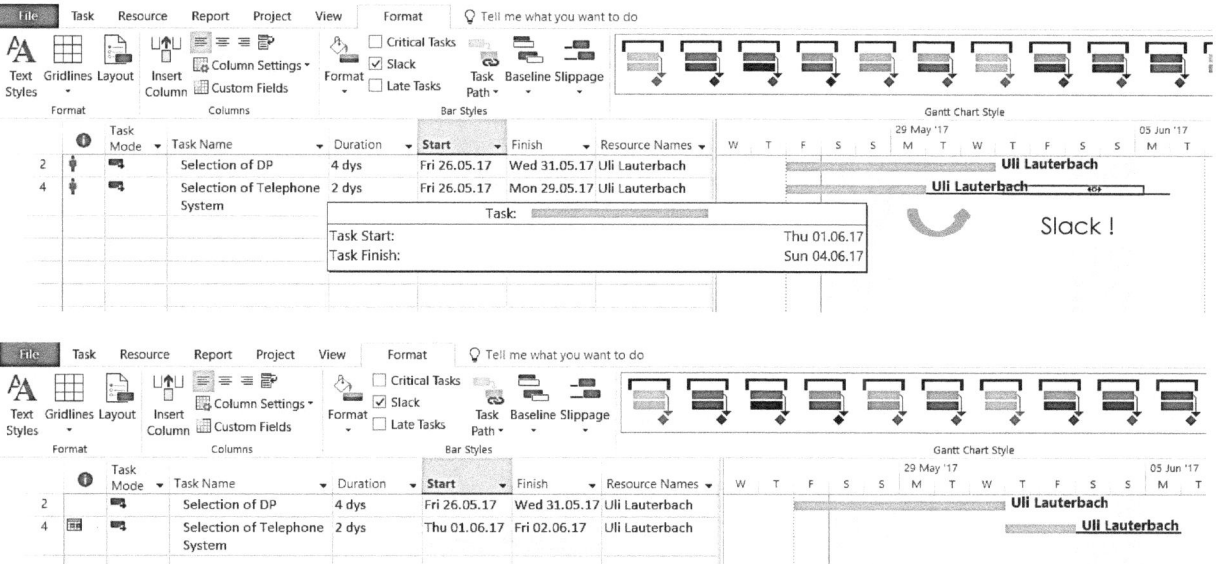

Unfortunately, it's not always that easy!!!

NOTES, COMMENTS:

8 COST MANAGEMENT

8.1 COST TYPES

Different cost types and cost rates can map many different requirements in Microsoft Project. The following cost variations can be mapped:

1. Work resources including costs based on actual effort (time-based)

In hours (standard rate per day, per week, per month). The costs are calculated depending on the usage duration: standard cost rate x duration (x assignment). Is also needed for material/machines that are charged by the hour.

2. Work resources including costs based on actual effort + one-time amount for costs per usage

As mentioned under 1: Additionally, one-time costs (once per usage/task), e.g. for the use of own tools, can be displayed, i.e. costs based on effort + costs per usage

3. Work resources including costs only per usage (task)

E.g. the craftsman or technician agreed on a fixed price, independent of the duration and hours of work

4. Material resources including costs based on consumption (with regards to measurement unit)

I.e. the costs are charged by consumed quantity, e.g. liters of fuel, cbm of sand, independent of the duration of the task

5. Work resources including costs based on consumption + one-time amount for costs per usage

As mentioned under 4. Additionally, one-time costs (once per usage/task) may occur. Water will be charged based on consumption in cbm but the fee for an additionally installed water meter has to be paid for only once.

6. Material costs including costs per usage

I.e. a one-time amount per task has to be paid for an expert opinion, independent of the task duration and of the effort for the expert opinion, alternatively as a work resource. In that case, the expert would charge his customer based on actual effort.

NOTES, COMMENTS:

7. Fixed costs (at task or at project level)

An input option to capture costs that cannot be assigned based on the above scheme, e.g. fixed costs that don't have to be described in detail such as insurance premiums for a construction work in a task. Fixed costs can be recorded per task or at project level (in the row "Project Summary Task"). Here, all insurance premiums can be inserted for the entire project. PLEASE NOTE: The costs are NOT totaled at project summary task level!!

8. Cost resources

A cost resource provides an opportunity to assign costs to a task by allocating a cost element (e.g. an airline ticket or accommodation) to a task. But the effective costs may vary in each task, such as e.g. travel expenses. This means that a cost resource doesn't depend on the workload of a task or of task duration.

9. Budget resources

This cost type allows you to enter budget values for work, material and costs at summary task level. These values can be compared with the accumulated values in the project. Such a comparison can be made without budget resources via custom fields.

NOTES, COMMENTS:

Below, we will show an example of a complete illustration of all cost types. Staff resources, fixed costs and variable travel expenses are considered, evaluated, contrasted and visually presented.

At first, the resources for different resource types are created in the mask "**View /Resource Sheet**". The selection is made in the column "**Type**". The costs for the cost resource won't be entered at this point as they can vary for each task and are instead directly recorded during the resource assignment (will be described below).

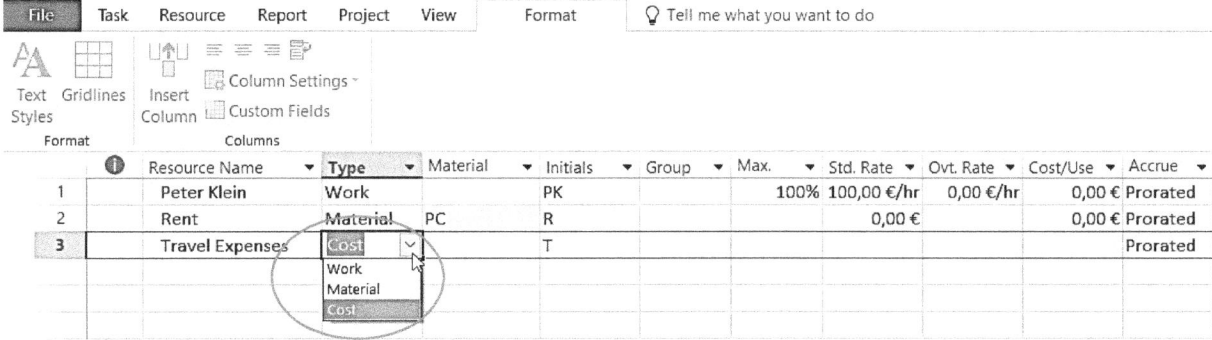

To assign the resources to the tasks, "View/Gantt Chart" is required. Go to resource assignment via the menu item "Resources", call up "Resource Assignment" and assign the resource per task as described in section 7.2.1.

The travel expenses are assigned to the causing task (here: Selection of Telephone System) and the amount (€300.00) is entered in the column "Costs" (mask "Assign Resources"). This way, travel expenses can be assigned to each task at different amounts. This cost presentation allows a definition of other cost types with a similar cost assignment/source.

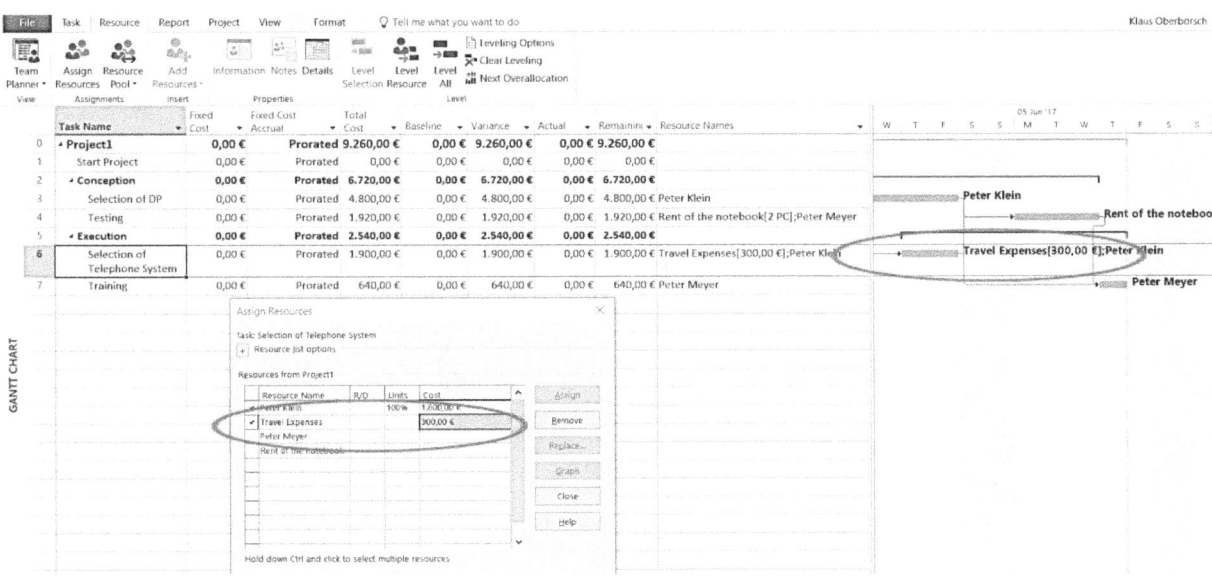

NOTES, COMMENTS:

The rent of the notebook is assigned to "Task Testing" as "one-time costs" = **costs per usage** (defined in the view "Resource Sheet"). Later, this amount will be shown in one amount as total costs including the resource costs.

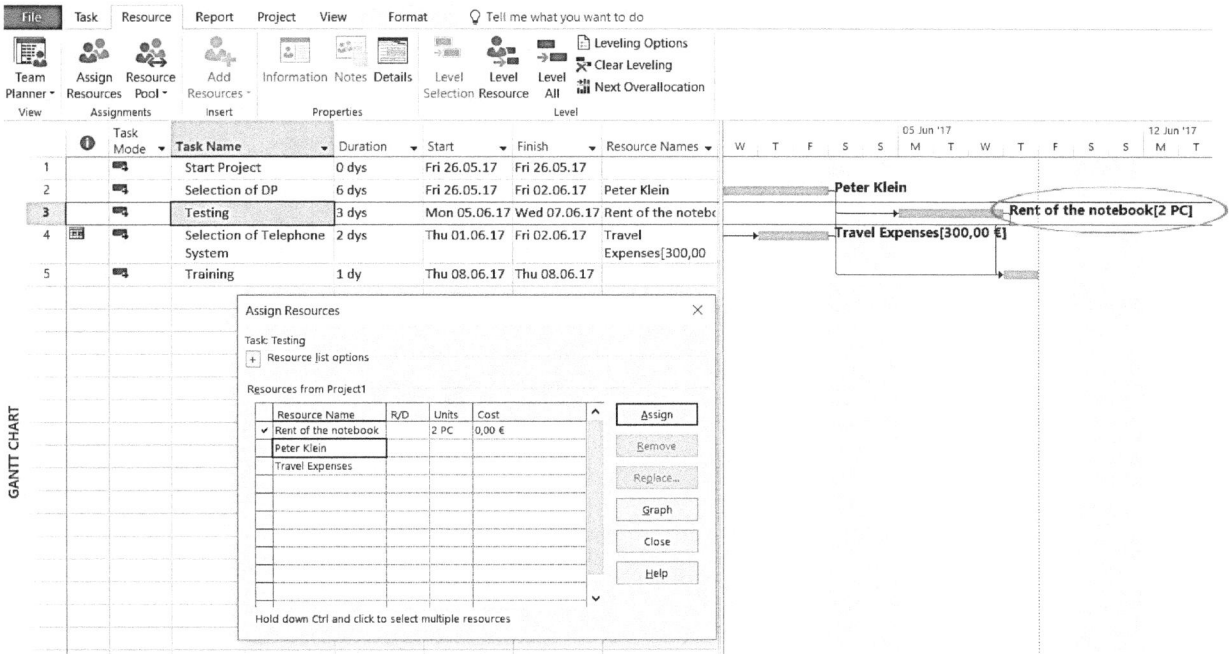

The costs are indicated per task using the menu item "View/Table/Costs". A subtotal is created. In case the project summary task has been selected via "Format/Project Summary Task", the totals are shown for the entire project as well.

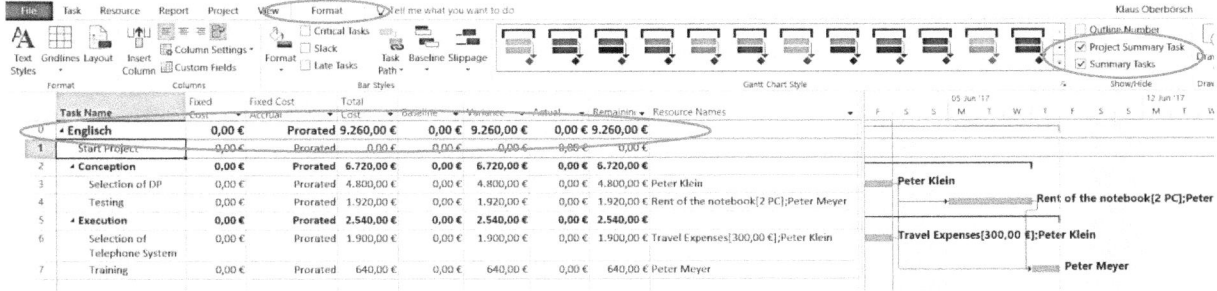

NOTES, COMMENTS:

The menu item "Report/Visual Reports" allows you to call up a visual representation of the cost types (in the template selection "All/Resource Cost Summary Report"). The minimum requirement is Excel 2013 SP2 (for other reports also min. Visio Professional 2013).

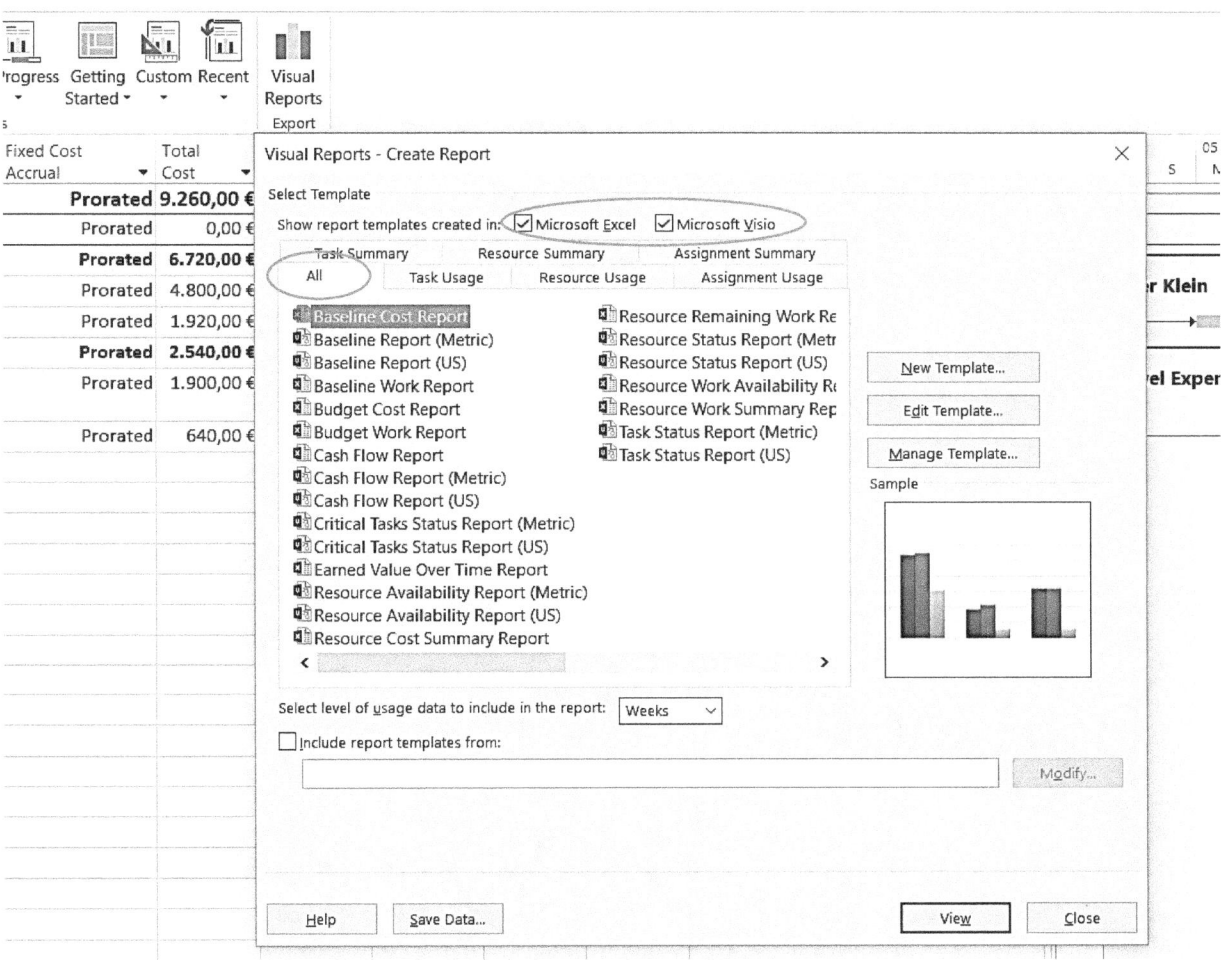

Excel allows to even format the <u>automatically</u> generated graphic with all available options and to show or hide data series/periods of time.

NOTES, COMMENTS:

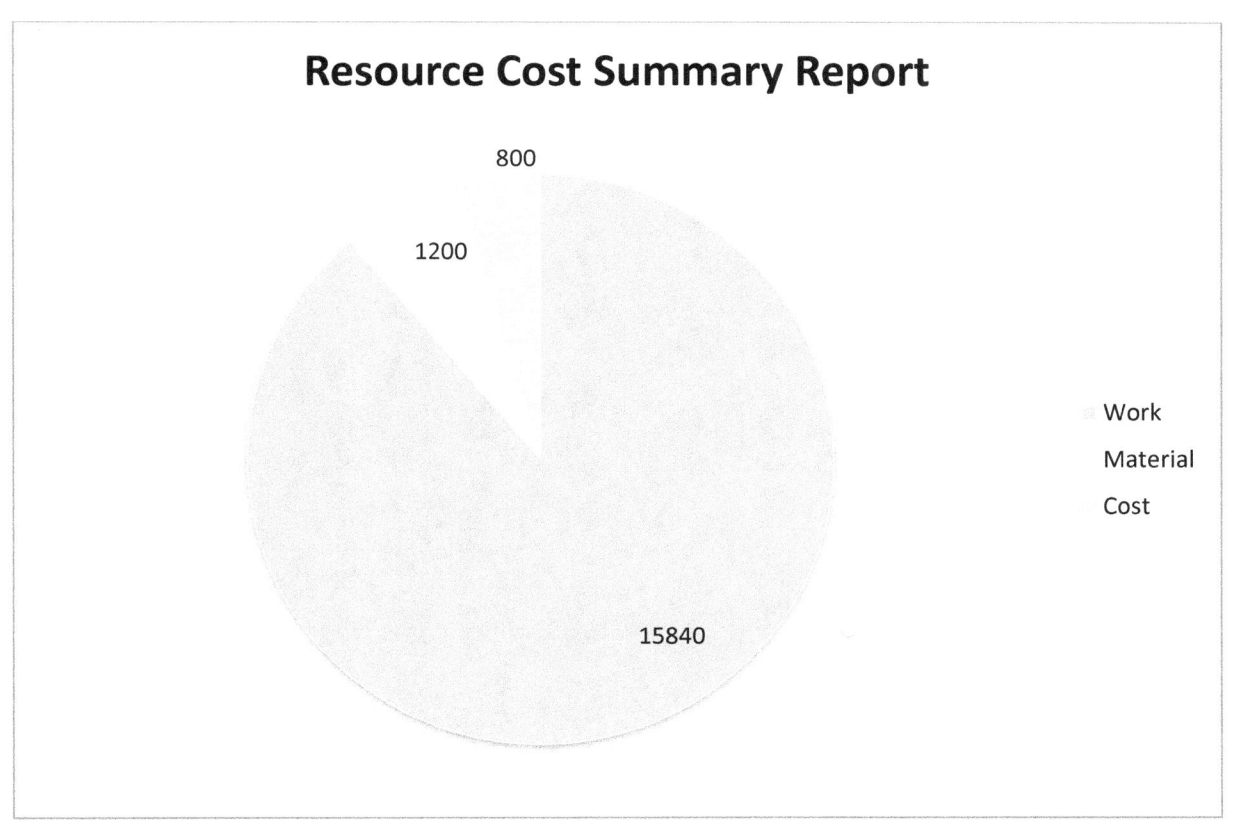

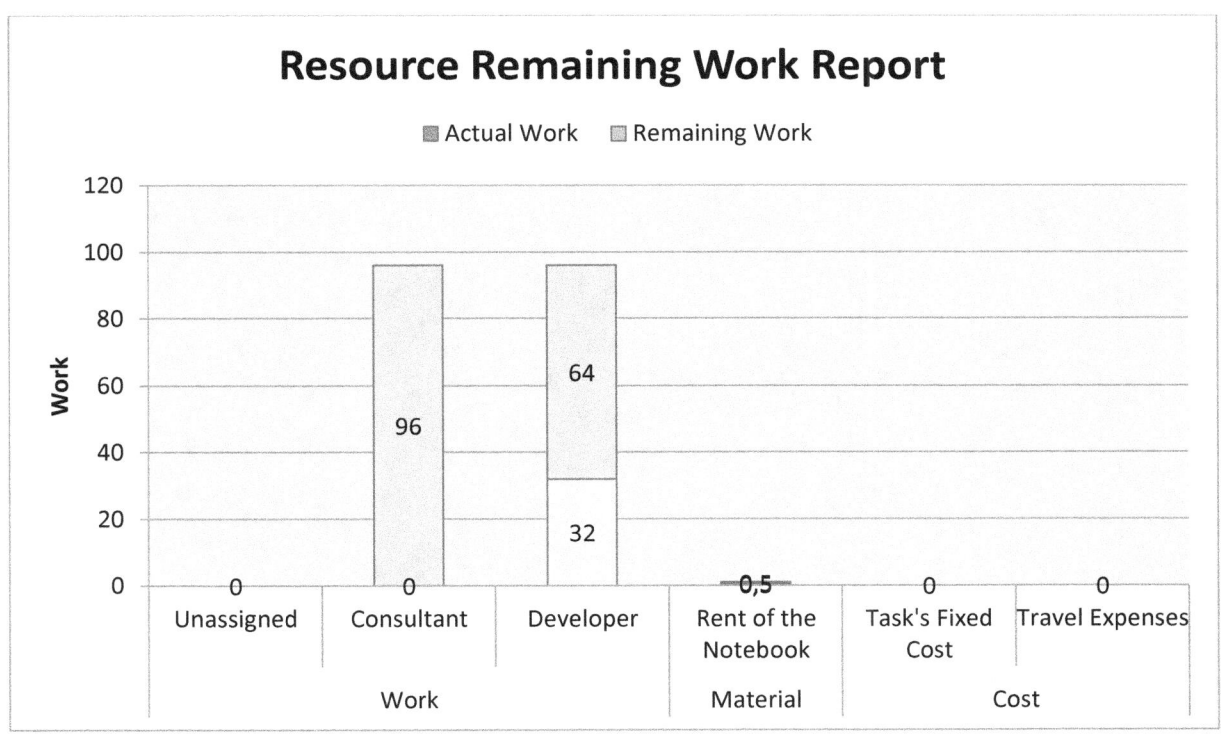

NOTES, COMMENTS:

8.2   BUDGET TRACKING

With its definition of budget resources, Microsoft Project offers the possibility to define costs and work budgets and to compare them with current values. This method enables the implementation but makes it also very complicated and complex. It's much easier to use custom fields, formulas and traffic light indicators.

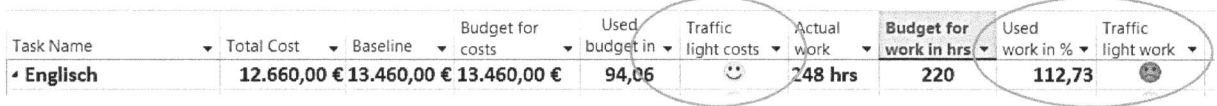

Task Name	Total Cost	Baseline	Budget for costs	Used budget in	Traffic light costs	Actual work	Budget for work in hrs	Used work in %	Traffic light work
◢ **Englisch**	12.660,00 €	13.460,00 €	13.460,00 €	94,06	☺	248 hrs	220	112,73	☹

Project summary tasks, detailed tasks were hidden.

In general, budget tracking should be displayed at project level. This requires setting up a so-called project summary task through the menu item "Format". Activate the summary task in the ribbon on the right. The summary task is always shown as the first task in a project and can't be moved!

For this presentation, some new custom fields were created and named accordingly. For example:

- **Budget for costs,** (the fixed budget is inserted for the project) can also be created from a dynamic link from Excel.
- **Used budget in %,** using the formula "IIf([budget for costs in €] >0;[costs]*100/[budget for costs in €] ;0)",Microsoft Project charges used cost budget based on current costs. The IIF query ensures that the calculation is only made in case of a recorded numerical value. Otherwise an error note appears.

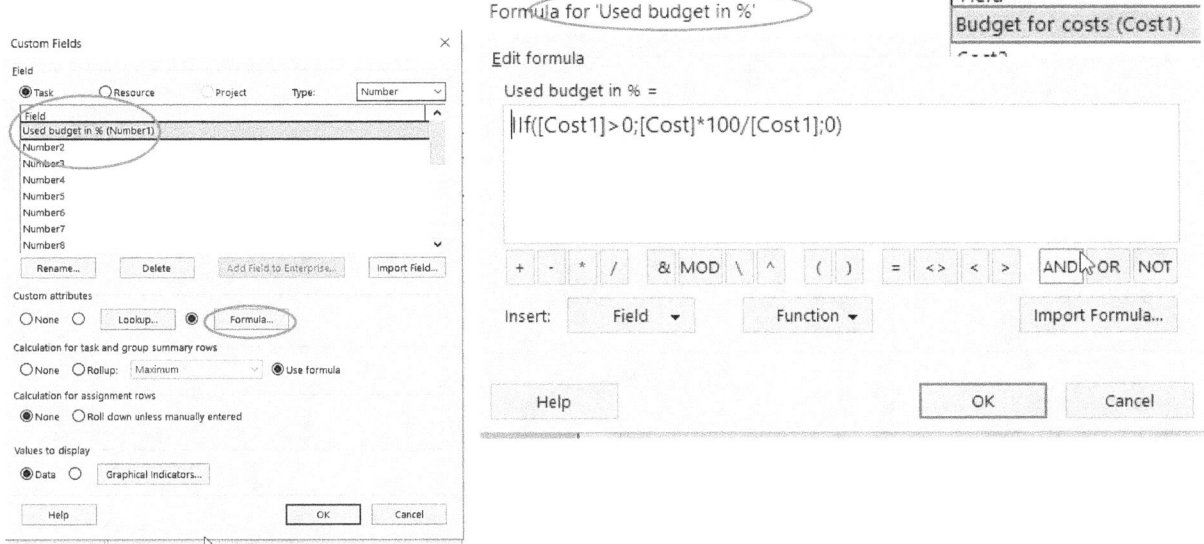

NOTES, COMMENTS:

- **Var. traffic light budget:** By means of the calculated value from "Used budget costs", data is displayed in the form of graphic symbols with defined traffic lights. The same formula as shown on the previous page is used here as well. However, in case of "Values to display", the function "Graphical Indicators" must be selected.

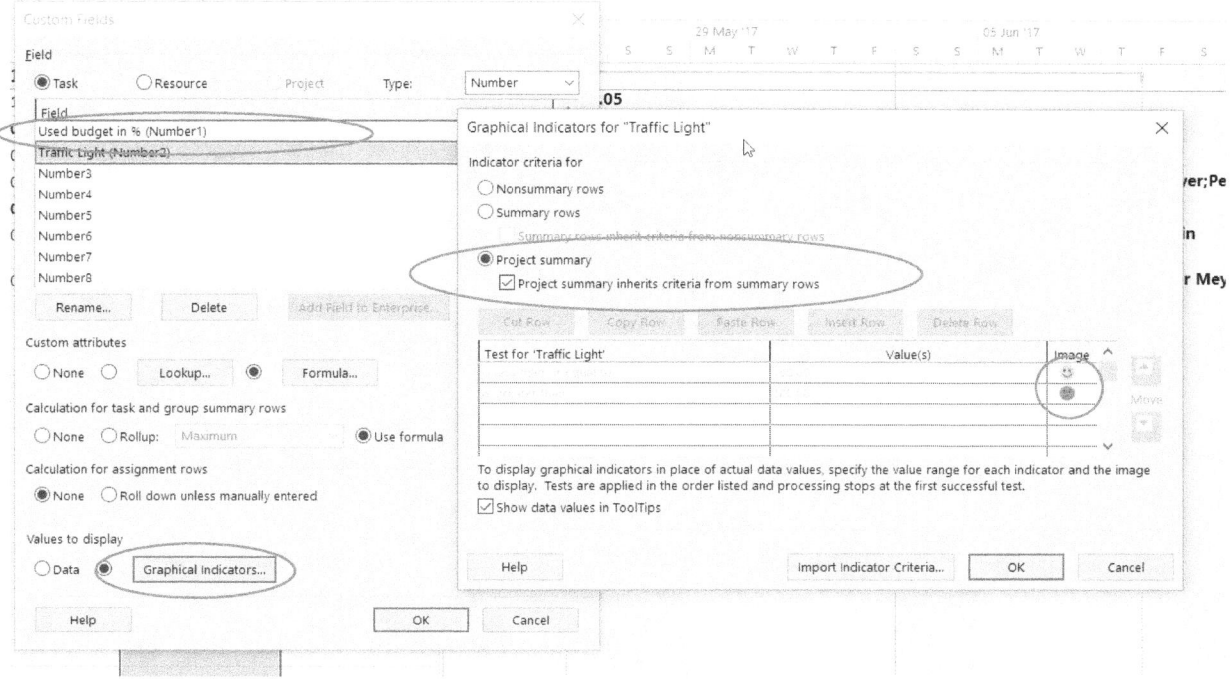

By all means, you should make sure you mark "Use formula" under "Calculation for task and group summary rows" as well a project summary task and the button Graphical Indicators – then select the underlying selection "Project summary inherits criteria from summary rows". In this example, smileys were used instead of colored traffic light symbols due to the black and white presentation.

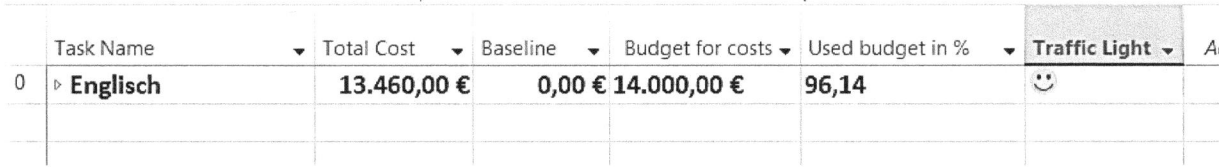

	Task Name	Total Cost	Baseline	Budget for costs	Used budget in %	Traffic Light	A
0	▷ **Englisch**	**13.460,00 €**	**0,00 €**	**14.000,00 €**	96,14	☺	

NOTES, COMMENTS:

- **Budget for costs in hours:** Here, the fixed work budget for the project has to be entered. It can also be generated with a dynamic link from Excel.
- **Used work in %:** Similar to the comparison of the cost budget, a formula for calculating the achieved percentage value is recorded in this field.
- **Var. traffic light work:** The same formula is used to show the variance of the work as traffic light function. The display through "Graphical Indicators" via a custom field is used instead of a pure data display.

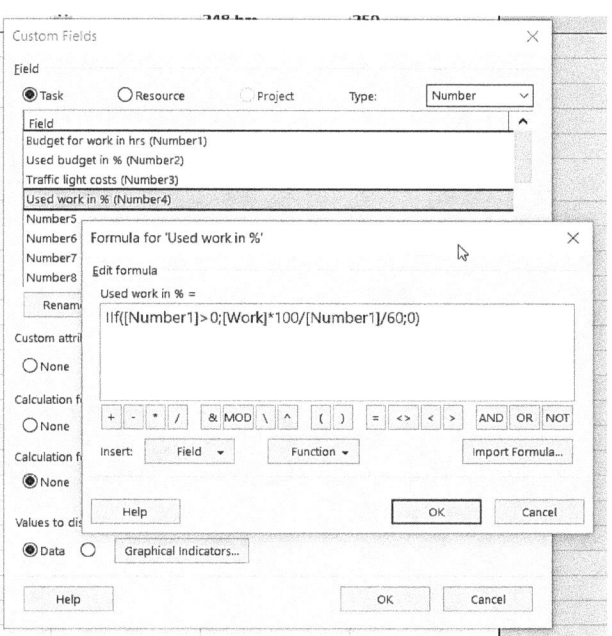

Now, the planned budget can directly be compared at project level with the actual incurred costs. There are almost no limits to calculations based on above formulas.

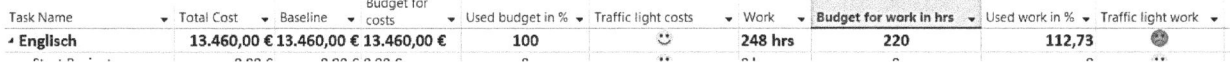

Task Name	Total Cost	Baseline	Budget for costs	Used budget in %	Traffic light costs	Work	Budget for work in hrs	Used work in %	Traffic light work
▲ Englisch	13.460,00 €	13.460,00 €	13.460,00 €	100	☺	248 hrs	220	112,73	☺

NOTES, COMMENTS:

9   PROJECT/TASK VIEWS

Information recorded in Microsoft Project can be specifically selected and displayed based on defined criteria. This function enables the user to control a current project and evaluate it/display it on a project-specific basis. The methods range from simple filter functions over groupings up to the use of variable fields with traffic light functions.

9.1   FILTER FUNCTIONS

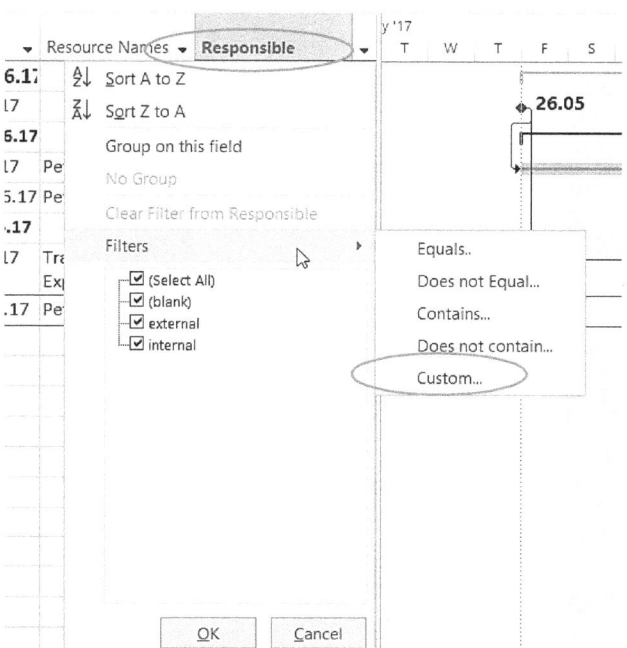

As you may already know from Microsoft Excel or Microsoft Access, the "AutoFilter" can also be activated in Microsoft Project per column using the small triangle next to the column title. Please see the below example of the field "Responsible" including the options "internal/external".

This function also allows you to activate the sort and group option based on this column.

Individual filters can be created using the filter option "Custom".

There are other pre-defined filters available in the menu item "View". Here, the "AutoFilters" are enabled or disabled, if needed.

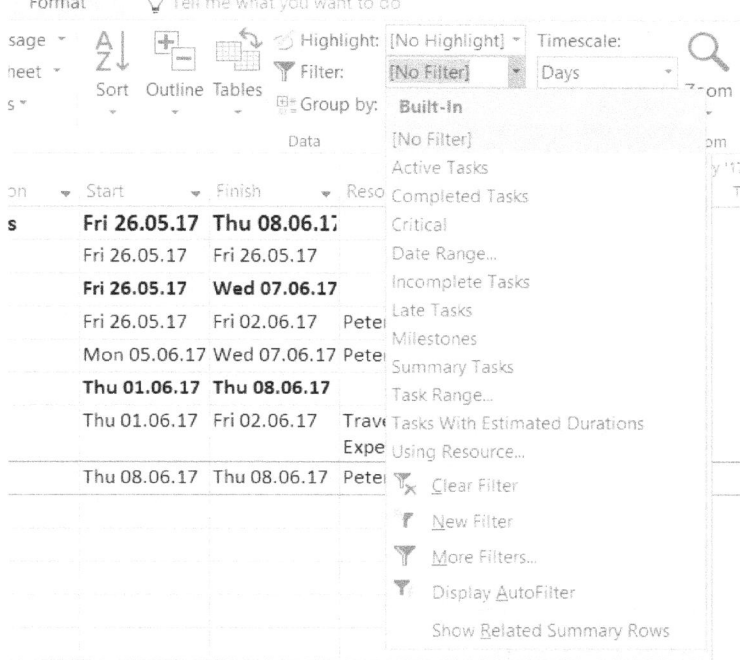

NOTES, COMMENTS:

9.2   GROUPING

The grouping of tasks, resources or grouping based on custom fields serves as a quick summary and clear presentation for project-specific views/evaluations.

The group function can be found under the menu item "View". The following example describes some available options.

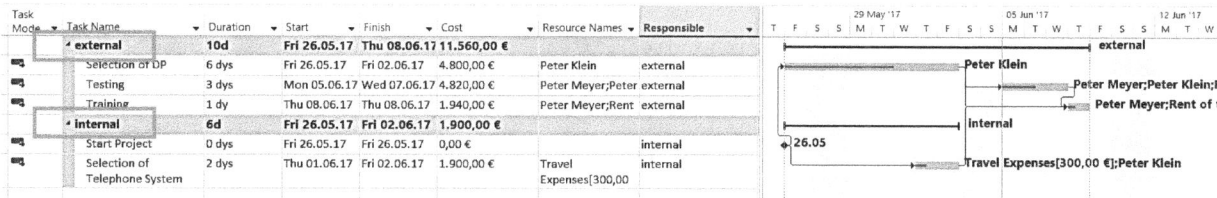

Single tasks are displayed in a selected group. Numeric fields like costs or work are totaled per group.

A grouping based on "Milestones" is also possible using the described filter function. But you can still see the links to the predecessors and successors. Here, the filter function may easier to use.

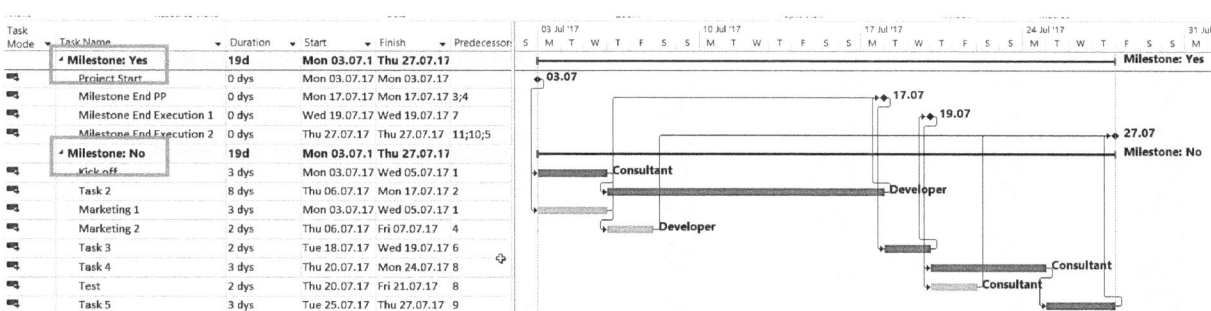

You can create your own groupings in the menu "View/Group by/More Groups". Or you use custom fields as a grouping element.

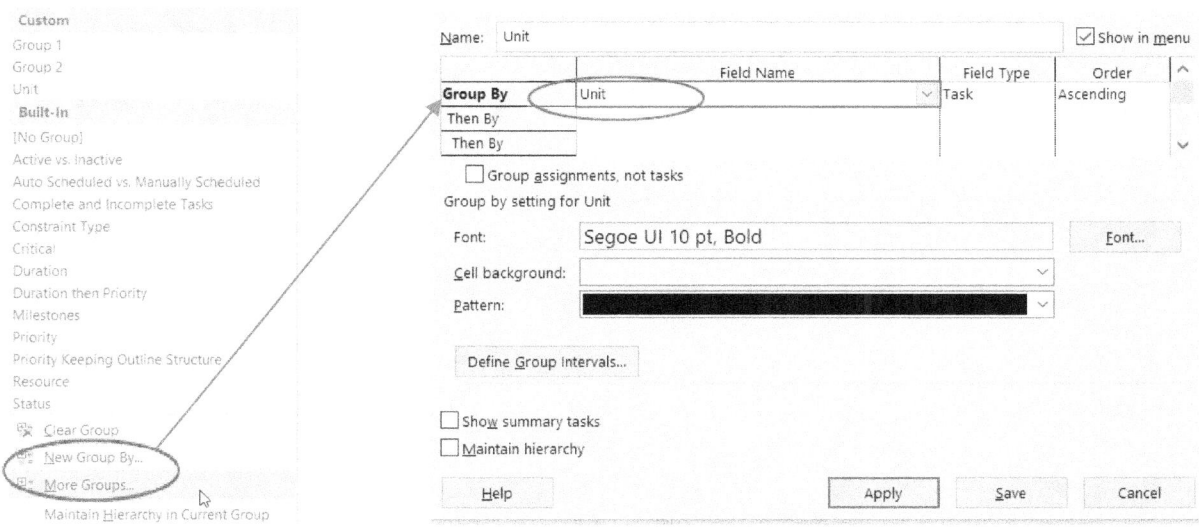

NOTES, COMMENTS:

9.3   CELL HIGHLIGHTING

The function "Cell Highlighting" offers a quick way to highlight special information. You can find it under the menu item "Format/Text Styles".

This way, you can also e.g. specially highlight tasks that are on the critical path.

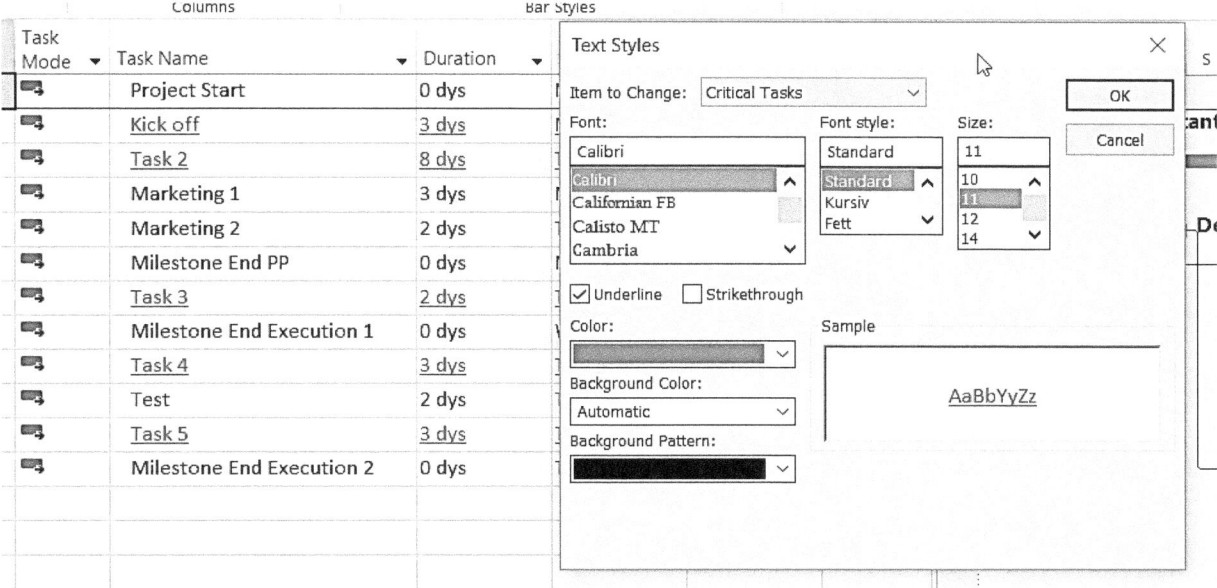

The entries to be changed can be selected from the drop-down menu. You can highlight cells by using a different font/font style and/or font size. Furthermore, the corresponding cells can be provided with a specific background color and/or a background pattern.

Together with a formatting of the bars, an individual Gantt chart including a display of the critical tasks in the table and calendar section can be created.

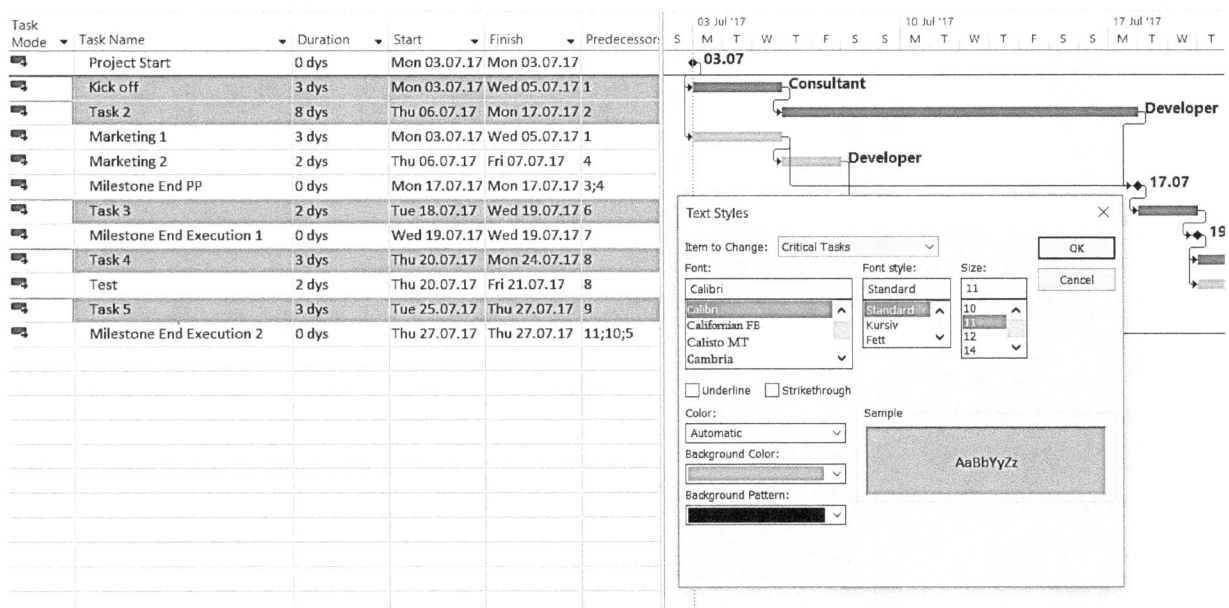

NOTES, COMMENTS:

10 PROJECT CONTROL/MONITORING

The tasks of a project manager monitoring a project include:

- Monitoring the progress of the project
- Entering current information on the project
- Creating a target/actual-comparison
- and using it to analyze changes in the project and their effect

10.1 SAVING BASELINE

First of all, the current schedule should be saved as a reference plan for a target/actual-comparison using "Set Baseline" under the menu item "Project". This is a requirement to reach a comparison with the original planning during the current project.

10.1.1 CAPTURING TARGET (SET BASELINE)

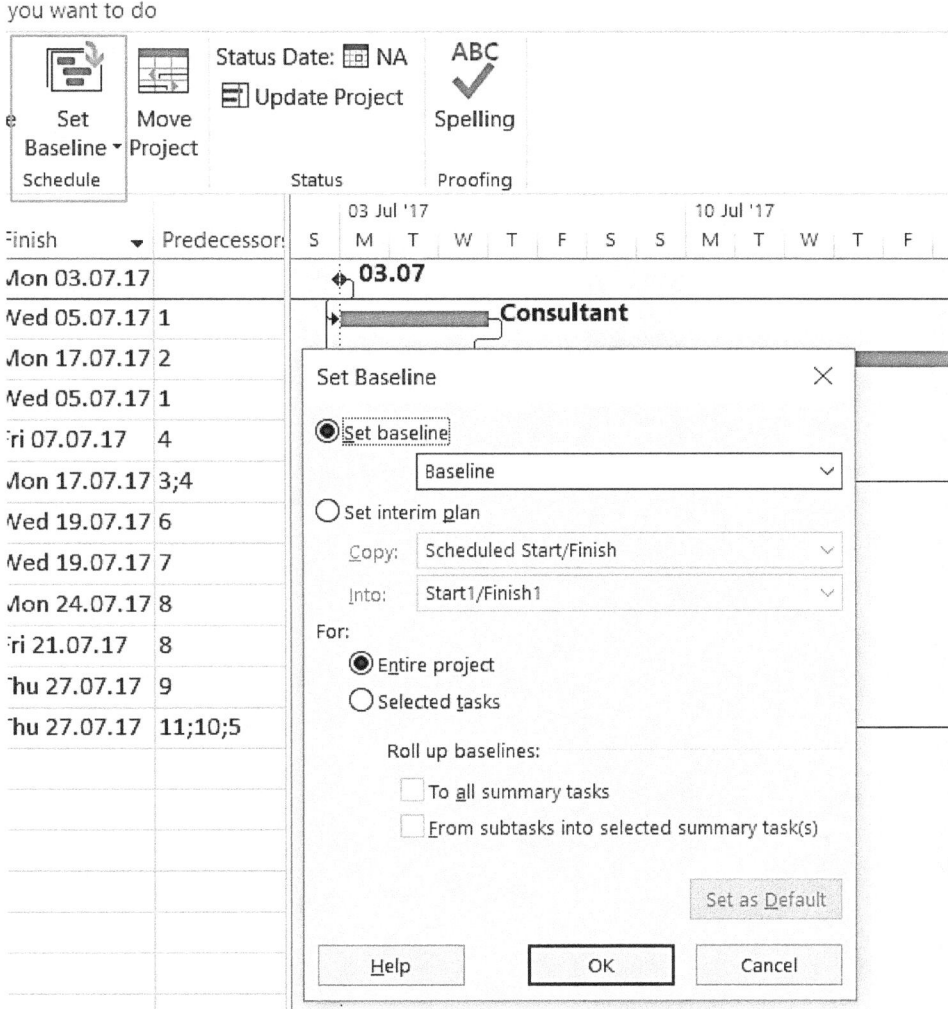

NOTES, COMMENTS:

You can save up to 11 baselines in total to allow full control over a longer period of time. The baseline can be saved for the entire project but also only for specific tasks. After saving the baseline, it will be added by the storage date.

When saving a baseline, Microsoft Project saves, among other things, the field contents from the fields "Baseline Start", "Baseline Finish" and "Duration".

10.1.2 COMPARE TARGET VALUES AGAINST ACTUAL VALUES IN A TABLE

The effects of saving the baseline become visual when the table **"Variance"** is called up in the view "Gantt Chart".

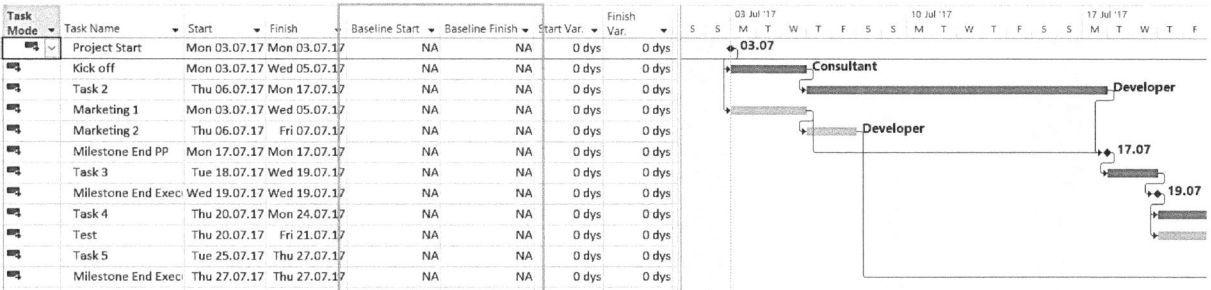

The fields **"Baseline Start"** and **"Baseline Finish"** are filled with the letters NA (Not Available) prior to saving the baseline.

After setting (saving) the baseline, you can first find the same values as in Start + Finish as no variances have occurred yet.

NOTES, COMMENTS:

In case the current project planning changes, e.g. the task duration, the table instantly shows variances. In the table below the task "Task 2" extends by 4 days. This doesn't result in a variance at the beginning but of course in the end due to a longer duration of this task. As a result, the column "Start Var." also shows 4 days. As this task is a critical task, the variance affects the project until its finish.

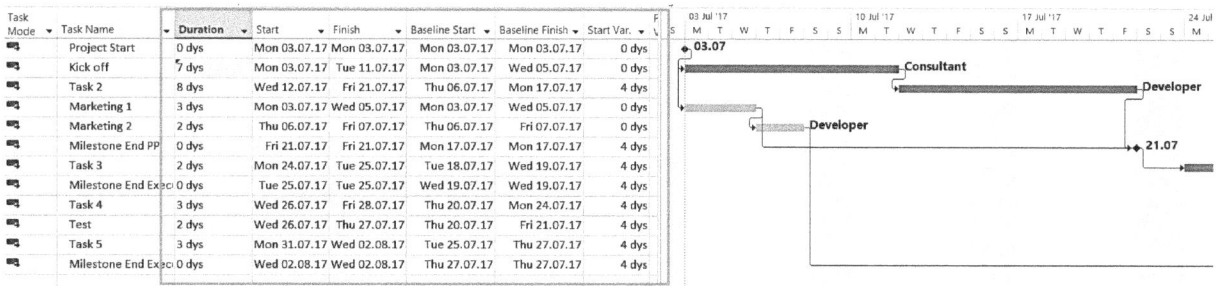

10.1.3 COMPARE TARGET AND ACTUAL VALUES VISUALLY

Of course, it's clearer and easier to visually present any postponements and their associated effects. There is an existing view available. In the view "Tracking Gantt", the target and actual values are prepared visually. The upper bars show the target plan, the lower bars show the actual plan. The postponement of milestones is also visible

In this example you can see that the task 2 takes longer than planned and will therefore finish later. This way, the next tasks will start with a delay and finish later as well.

NOTES, COMMENTS:

10.1.4 "CHANGING"/CLEARING BASELINE

When the target values have been saved too early and the baseline changes, it can be deleted or overwritten.

Clearing baseline

The baseline/s can be cleared in the menu item "**Project/Set Baseline**".

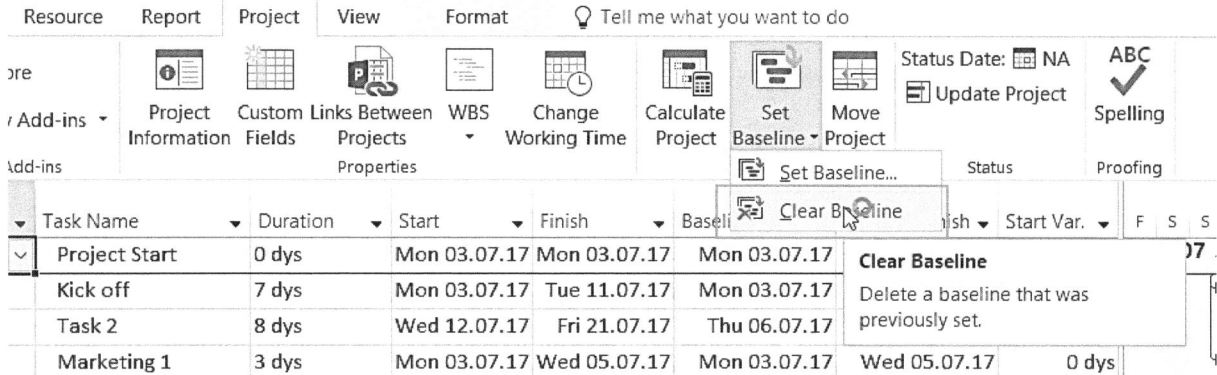

Changing baseline

A change in the proper sense cannot be made. If changes in the plan have been made, current data is saved as a new baseline after prior confirmation.

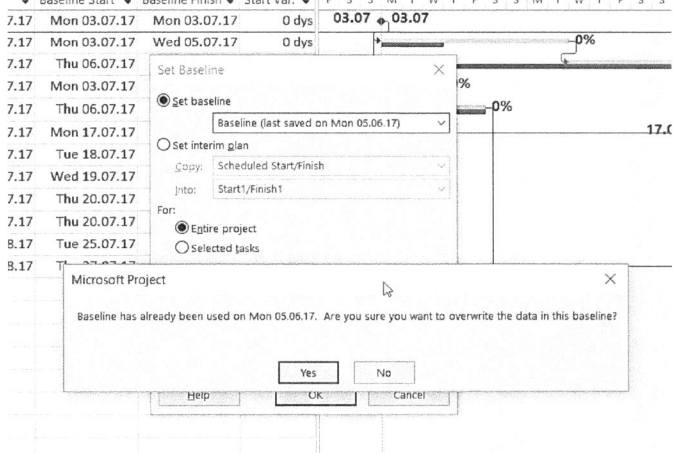

NOTES, COMMENTS:

10.2 PROJECT CONTINUATION

After the project has been saved to enable an actual/target-comparison, updates are made in the course of the project.

The progress in Microsoft Project is recorded through the performed work – either by the hour or in % (percent). The view "Table/Work" is ideal for this.

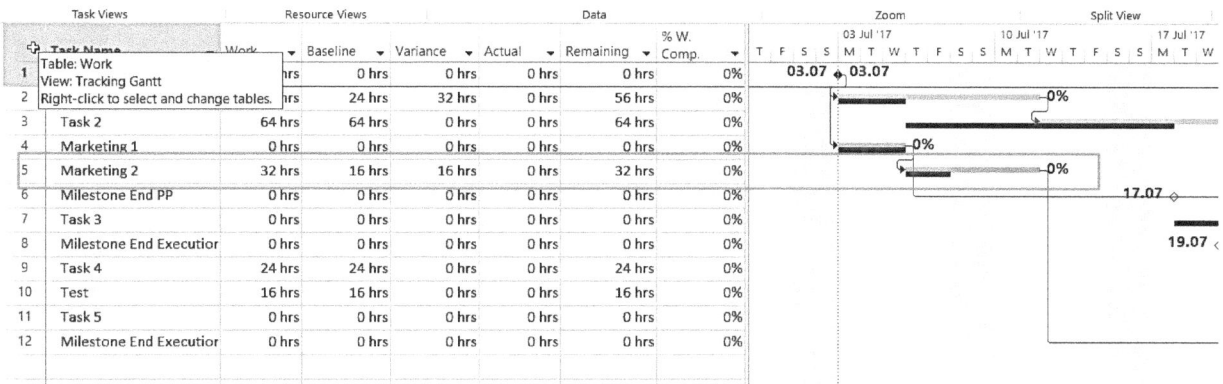

Due to the saved baseline (column "Baseline"), variances can immediately be identified. The performed working hours can be inserted into the column "Actual" or, alternatively, the degree of completion in % that, in turn, is converted to hours.

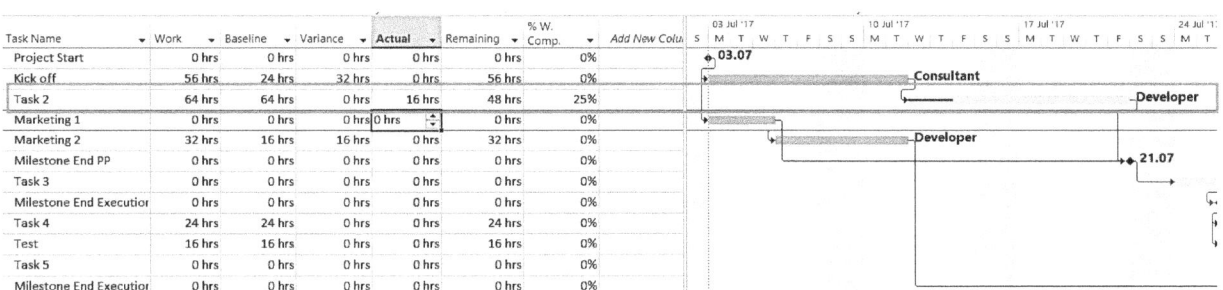

Entering 16 hours in the column "Actual" results in a completion of the task of 25%. At the same time, a black progress line is displayed in the Gantt chart that presents the progress visually.

NOTES, COMMENTS:

10.3 EVALUATING MONITORING INFORMATION

The integrated reports in the menu item "Report" are suitable for monitoring the project. It offers a great selection of predefined reports on tasks, costs, work and much more.

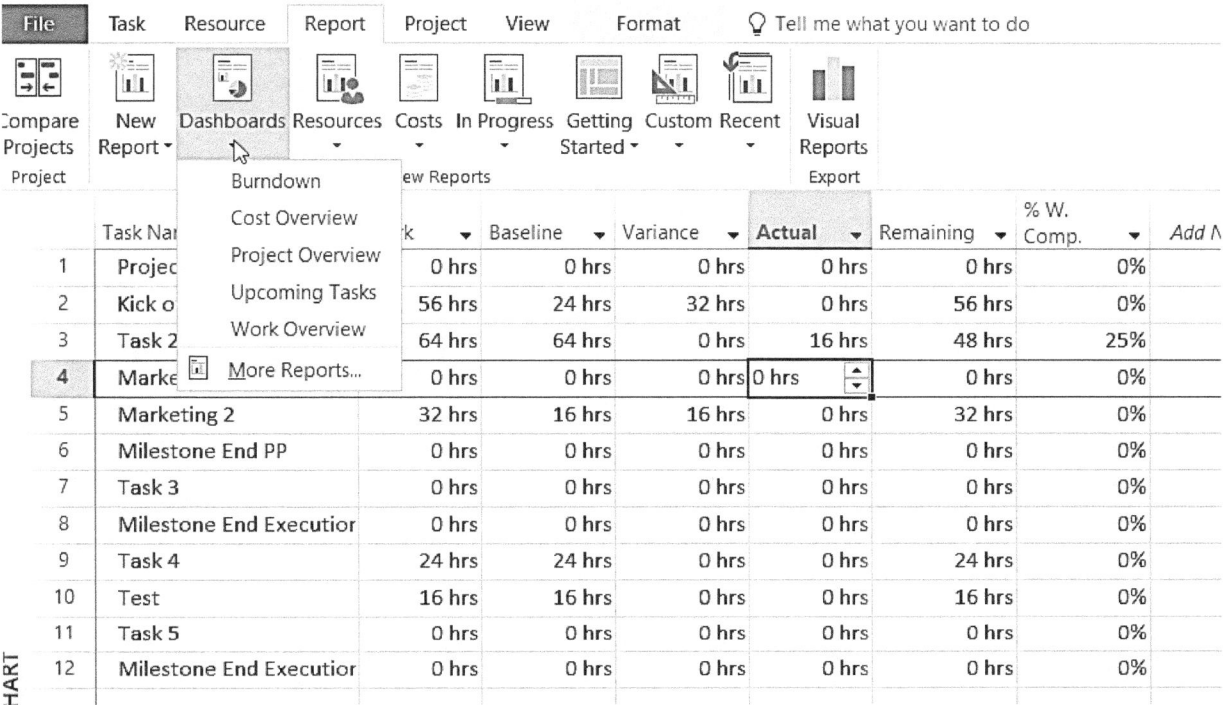

Below, you will find an example for possible report types, in this case "Project Overview". For a more detailed description of reporting, please refer to section 13.

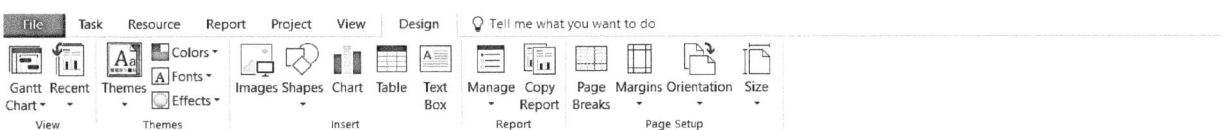

PROJECT OVERVIEW

MON 03.07.17- WED 02.08.17

% COMPLETE

6%

MILESTONES DUE
Milestones that are coming soon.

Name	Finish
Project Start	Mon 03.07.17
Milestone End PP	Fri 21.07.17
Milestone End Execution 1	Tue 25.07.17
Milestone End Execution 2	Wed 02.08.17

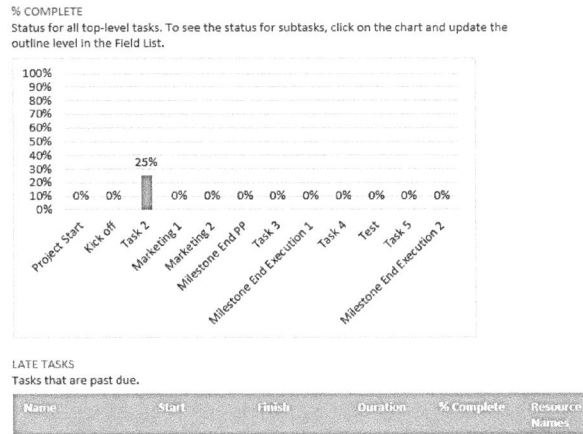

NOTES, COMMENTS:

11 CUSTOM FIELDS

Microsoft Project includes custom-fields that enable you to save individual information in fields and evaluate them through filters or the grouping function.

Custom fields are perfectly suitable to calculate and prepare specific data using formulas for specific projects. For this, e.g. the fields "Text 1-30" and "Number 1-20" are available.

Different examples for using custom fields will be described below.

11.1 LOOKUP FIELDS

It will often be required to map assignments, responsibilities, cost centers per task tin order to filter, group or evaluate a project view based on these additional information.

E.g., you would like to map in a column if internal or external units are responsible for a task. By inserting a custom field (here: "Text 1") and renaming this field, you have an entry option. However, the quality of entries influences the quality of evaluations. The entry may vary depending on the user. Possible entries could be internal/external. But also int./ext. could be possible or just i/e.

To unify entries or restrict them to specific values, lookup fields are set like in Excel. Select **"Custom fields/Add New Column"**. Select the field "Text 1" and click on button "Rename Field". Enter a descriptive name (Unit) so that you can find the field later.

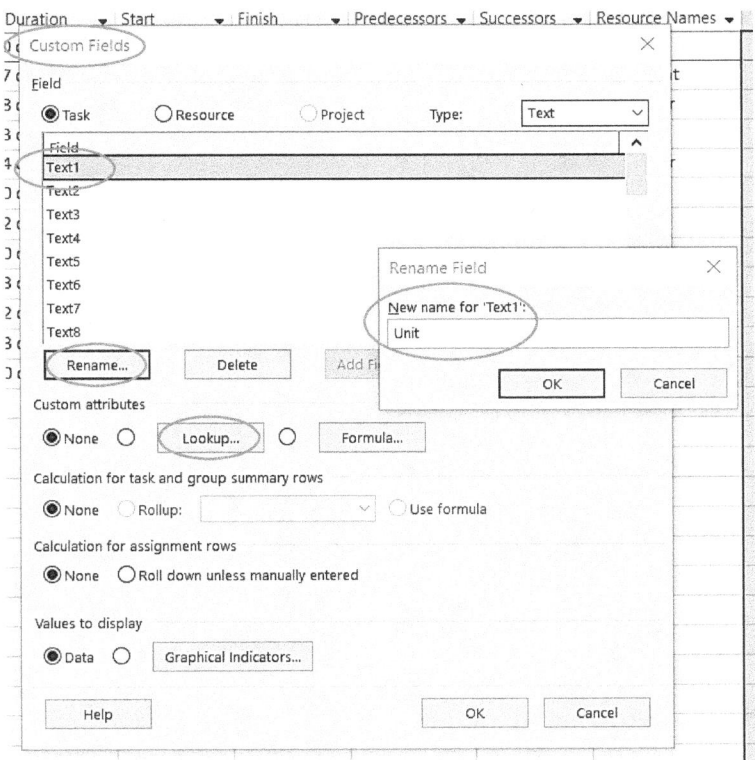 Save under the new name by clicking OK. Then, click on the Lookup button. All values that can be selected later should be entered along with a help/information text.

NOTES, COMMENTS:

Alternatively, lookup values can also be imported (e.g. from Excel) or inserted via copy and paste. The sequence can be changed using the function "Move".

By clicking on "Close", the entry is completed, and the OK button is clicked for confirmation in the mask "Custom Fields". Now, this field must be integrated using "Add New Column" in the table. The field can be found under "Text1" as well as under the newly assigned name.

In this field you can select only the displayed values ensuring data integrity.

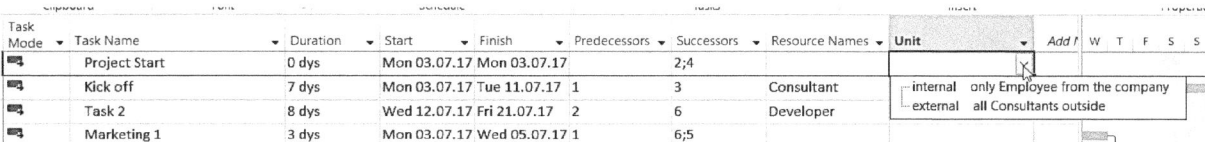

Task Mode ▾	Task Name ▾	Duration ▾	Start ▾	Finish ▾	Predecessors ▾	Successors ▾	Resource Names ▾	Unit ▾	Add	W	T	F	S	S
🖥	Project Start	0 dys	Mon 03.07.17	Mon 03.07.17		2;4								
🖥	Kick off	7 dys	Mon 03.07.17	Tue 11.07.17	1	3	Consultant	internal — only Employee from the company external — all Consultants outside						
🖥	Task 2	8 dys	Wed 12.07.17	Fri 21.07.17	2	6	Developer							
🖥	Marketing 1	3 dys	Mon 03.07.17	Wed 05.07.17	1	6;5								

These fields can be used for filter functions or groupings. The filter function in the column header is a very easy way to do so. It opens with one mouse click on the black triangle. The same applies to the grouping function. The custom field is selected as a change of group.

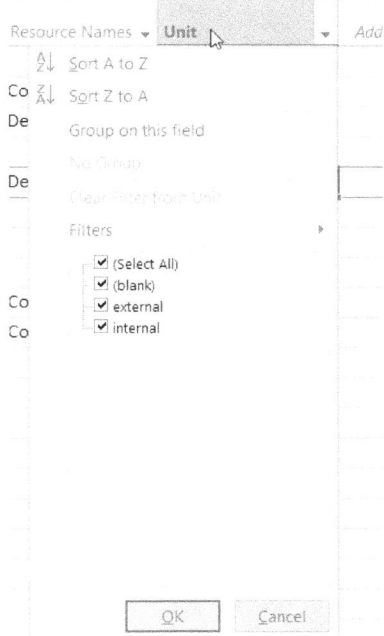

NOTES, COMMENTS:

In numeric fields, such as costs or work, a total is calculated separately for each group. The individual grouping can be saved under one name and be displayed in the menu "Group By".

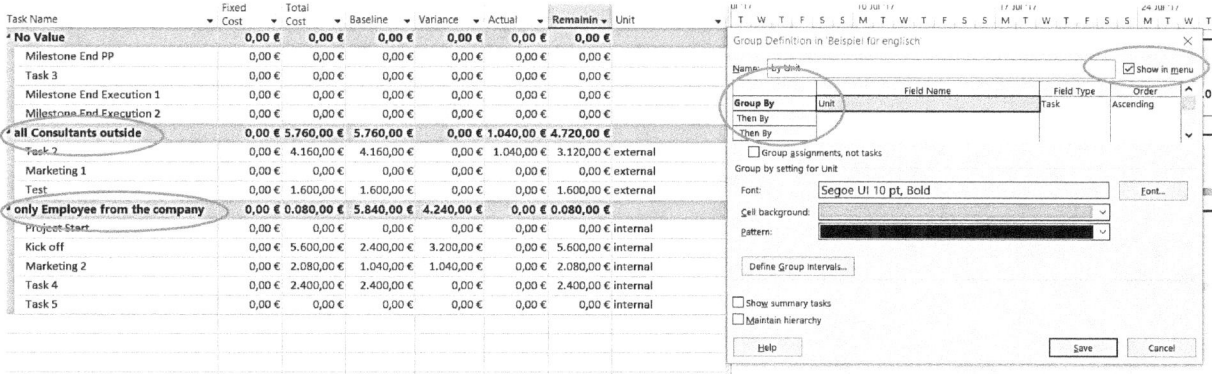

NOTES, COMMENTS:

12 MULTI-PROJECT MANAGEMENT

The standalone version allows to receive a product portfolio for the overview of multiple projects. This can be made for single independent projects but also for big projects that have been divided into subprojects and are therefore planned by the responsible subproject manager.

All functions described so far can also be used within the framework of multi-project management, incl. custom fields, formulas, traffic light functions and cost overviews.

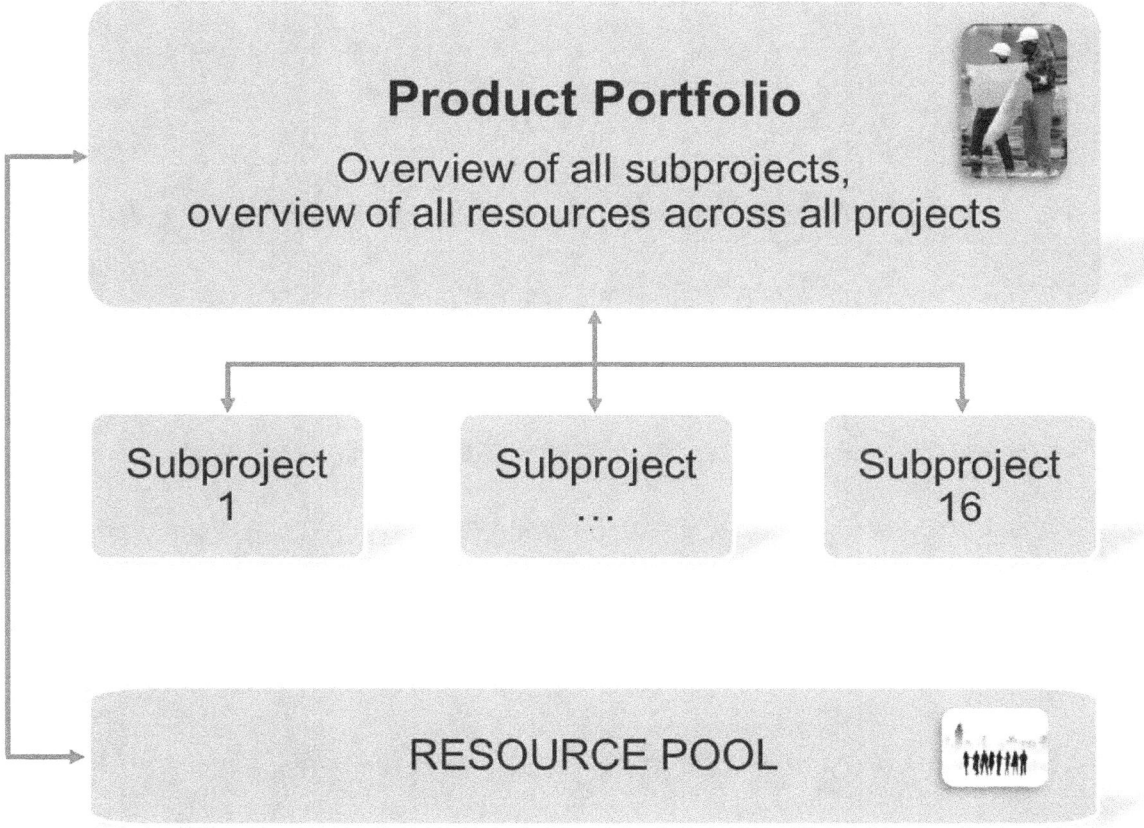

A resource pool is created for resource scheduling in which all available resources for all project participants are included. Each "subproject" only uses the resources defined in the resource pool. This way, a comprehensive overview of all resources regarding availability, work per project, costs per project and the like can be displayed in the project portfolio.

NOTES, COMMENTS:

12.1 SUBPROJECTS

A project portfolio (overview of all projects/subprojects) can be created from existing Microsoft Project files or from projects to be newly created. The process is the same.

The desired projects are selected through the menu item "Project/Subproject". Using the Ctrl button, you can mark multiple projects at the same time and insert them.

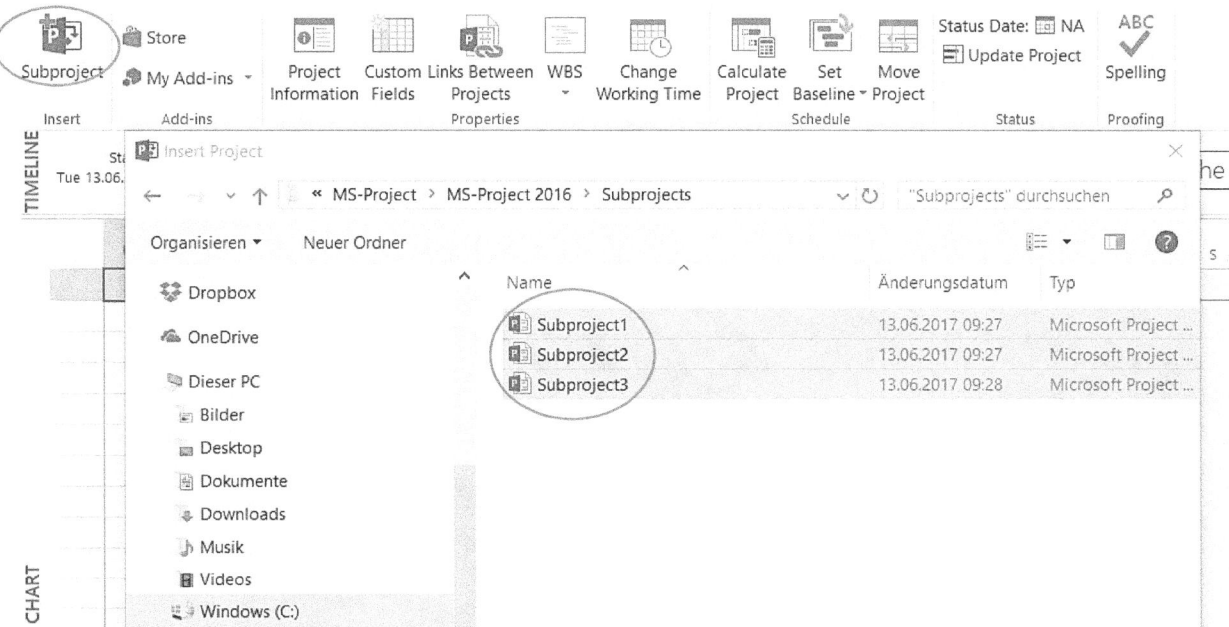

The inserted projects/subprojects are displayed like summary tasks, the respective file names are used as project header. Depending on the complexity, the detailed tasks can be shown or hidden. The sequence of the inserted projects can be changed by moving a specific project with the left mouse button to the desired position.

NOTES, COMMENTS:

In this example, the resources from the resource pool have already been assigned and an overallocation across all projects is visible.

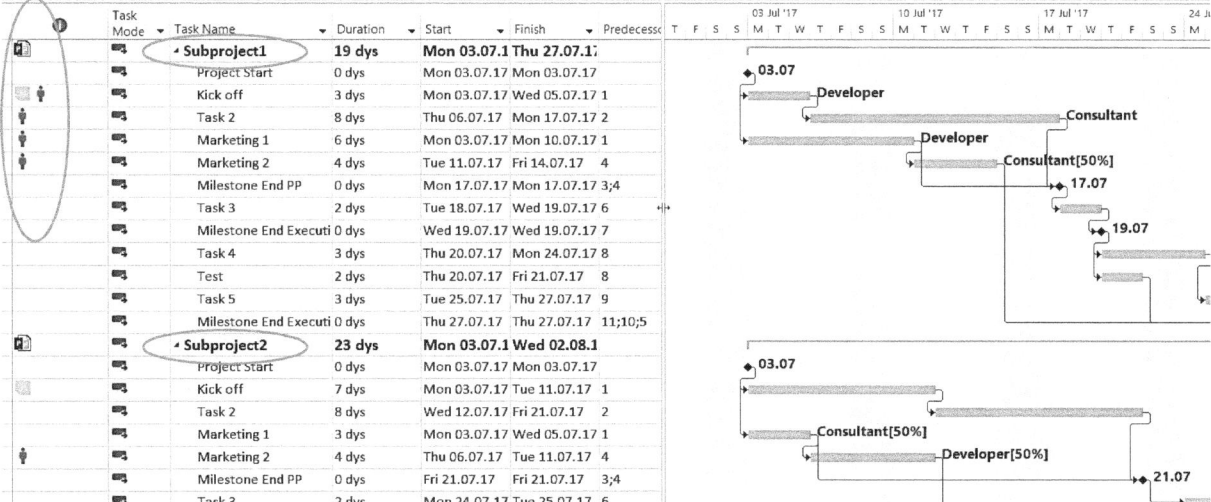

Similar to a single project, cross-project task relationships can be outlined here. "Task 2" from "Subproject1" is the predecessor of the task "Marketing 2" in "Subproject 2". In each project, these cross-project task relationships, which are specially formatted, are displayed as external successors/predecessors

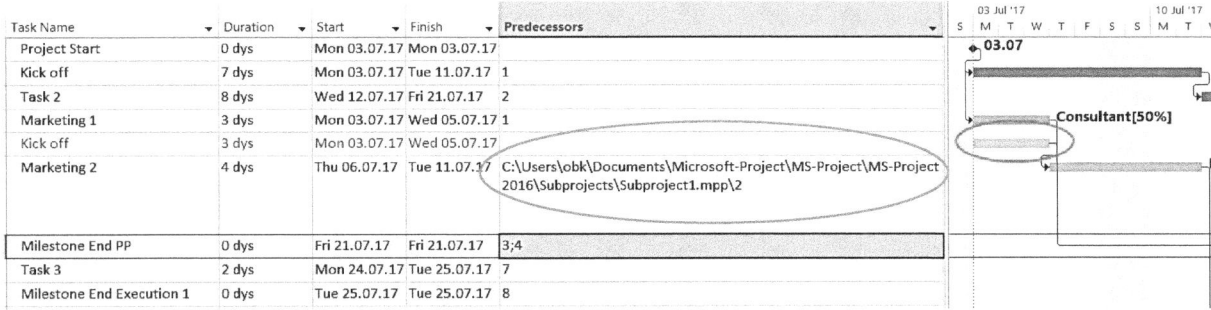

View in Subproject 2

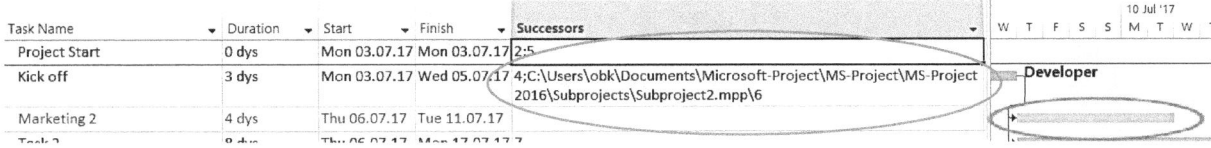

View in Subproject 1

NOTES, COMMENTS:

12.2 CREATE RESOURCE POOL

To receive a complete overview of all resources across all projects, resources scheduled in individual projects must be inserted into a resource pool and scheduled from there beforehand.

The resource pool is created in the respective master project/project portfolio. The procedure corresponds to the entry of resources under section 7 with all possibilities of different resource types.

When planning resources in individual projects, you must create a connection to the resource pool using the menu item "Resources/Share Resources" prior to the actual resource assignment. For this purpose, the project that was created in the resource pool, must be open in the screenshot "Project Portfolio".

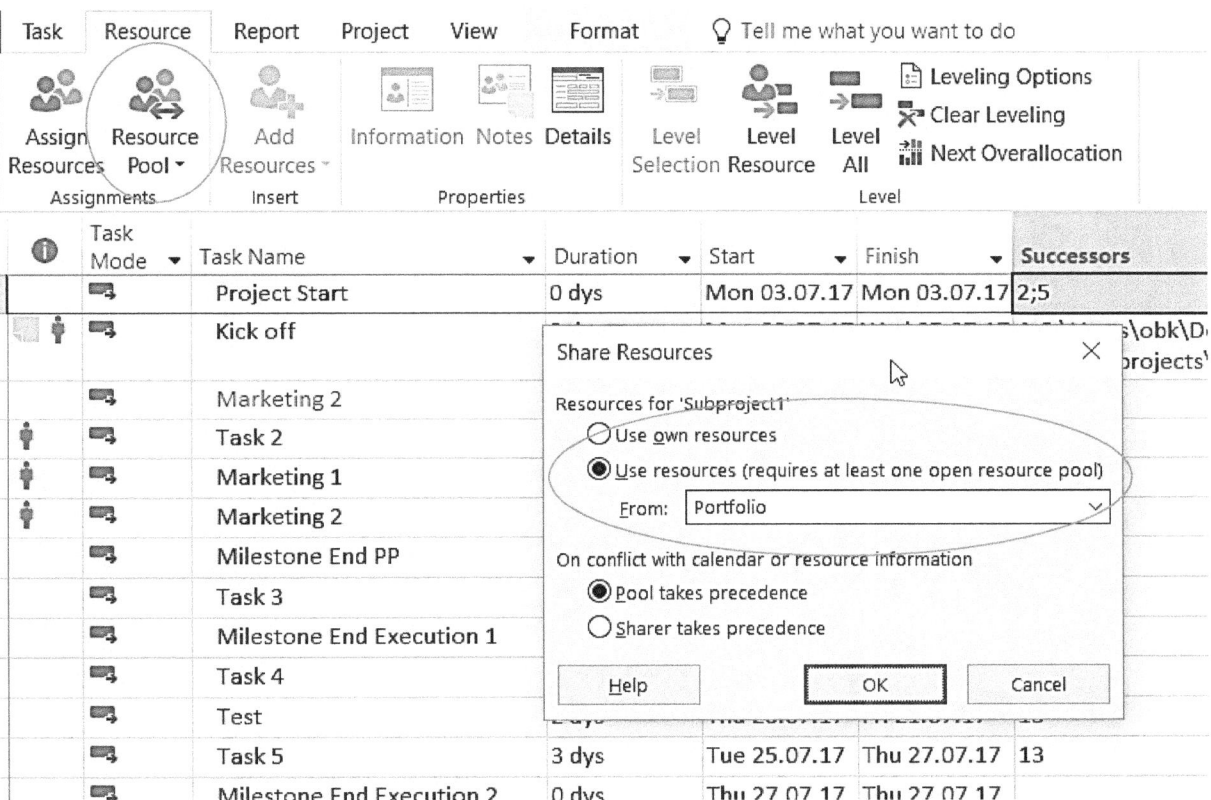

Then, the resources are displayed in the table and can be assigned to each tasks as described in section 7. Please note: When saving a project always answer the query "Update Resource Pool" with "Yes" and update changes in all projects.

NOTES, COMMENTS:

12.3 PROJECT PORTFOLIO/OVERVIEW

After the integration of all subprojects in the master project (here: project portfolio) and the assignment of the resources of each project via the resource pool, a cross-project presentation is now available.

All tables and views from the normal project planning can also be used here. For the presentation of totals (costs, work), the function "Project Summary Task" under the menu item "Format" should be enabled. Each project is totaled, and a total is created in the first row.

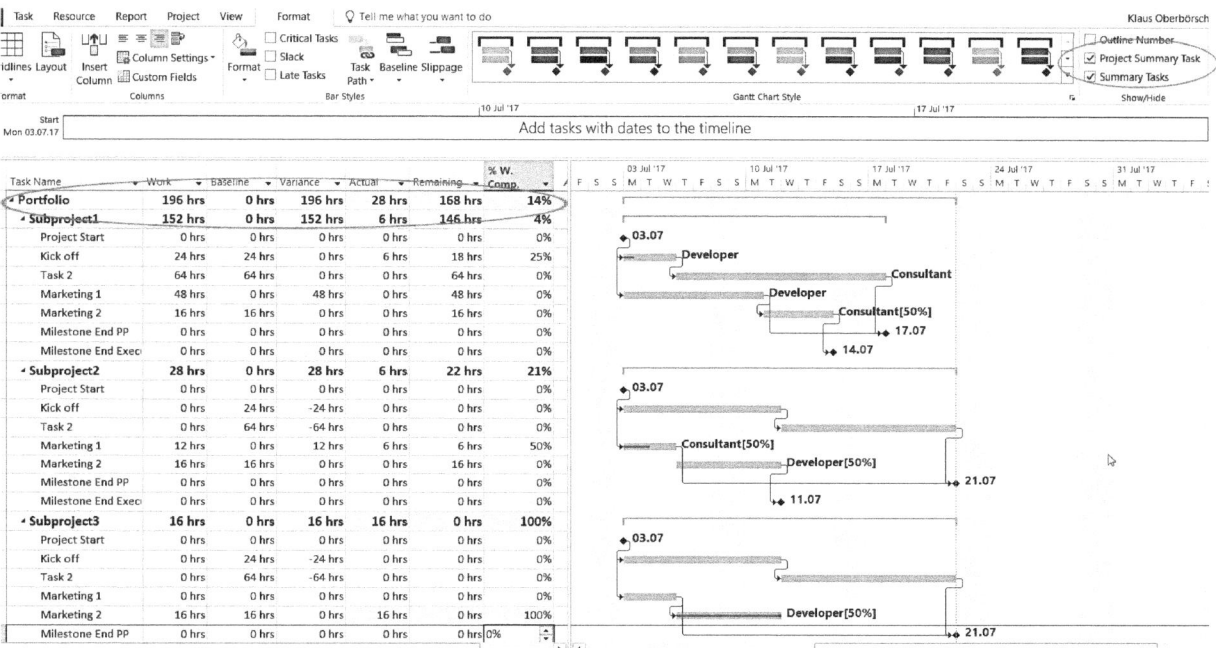

The usage of each resource per subproject can be seen in the view "Resource Usage". The menu item "Insert Column/Project" will be displayed as well.

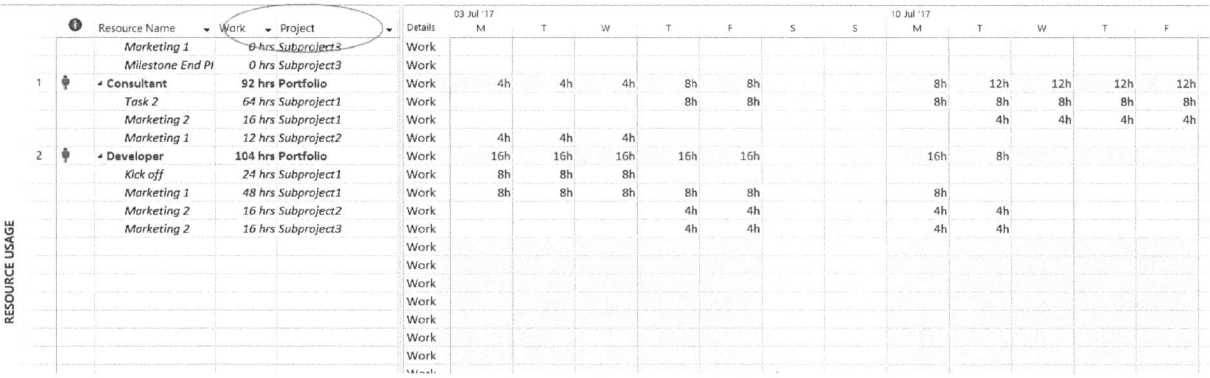

NOTES, COMMENTS:

13 REPORTS AND GRAPHIC EVALUATIONS

13.1   VISUAL REPORTS

The visual reports templates are shown in the toolbar "Reports" and are divided into six categories. The visual reports of each category are described in the following sections.

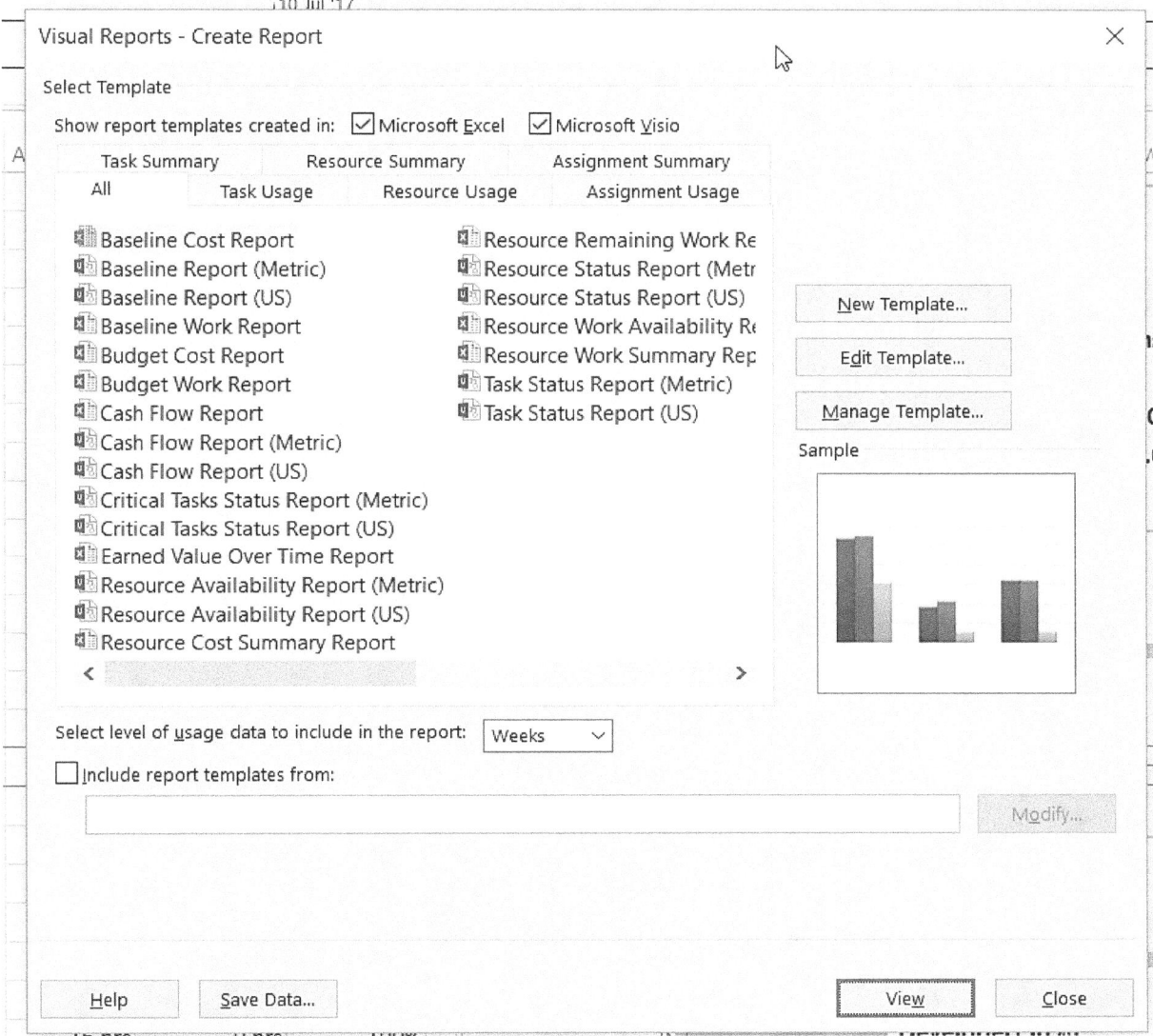

To ensure that all report templates can actually be used, the programs Microsoft Excel and Microsoft Visio Professional must be installed.

NOTES, COMMENTS:

13.2 OVERVIEW OF VISUAL REPORTS

As described in the beginning of this chapter, there is a variety of graphic report templates available which can be modified specific application programs.

13.2.1 CATEGORY "TASK USAGE"

Name	Type	Description
Cash Flow Report	Excel	This report allows you to display a Gantt chart with values for costs and cumulated costs over a specific period of time.

13.2.2 CATEGORY "RESOURCE USAGE"

Name	Type	Description
Cashflow Report	Visio	This report compares the originally planned work and costs with scheduled work and scheduled costs. Indicators show when the originally scheduled work exceeds the scheduled work and when the originally scheduled costs exceeds the scheduled costs.
Resource Availability Report	Visio	This report can be used to display a Gantt chart illustrating the work and the remaining availability of the resources of the project, broken down by resource types (work, material and costs). A red mark is shown next to the overallocated resources.

NOTES, COMMENTS:

Resource Cost Summary Report	Excel	This report can be used to display a pie chart illustrating the assignment of the resource costs under the three resource types: **costs**, **material** and **work**.
Resource Work Availability Report	Excel	This report can be used to display a Gantt chart illustrating the overall capacity, work and remaining availability for work resources over a specific period of time.
Resource Work Summary Report	Excel	This report can be used to display a Gantt chart illustrating the overall capacity of the resources, the work, the remaining availability as well as the actual work in work units

13.2.3 CATEGORY "ASSIGNMENT USAGE"

Name	Type	Description
Baseline Cost Report	Excel	This report can be used to display a Gantt chart illustrating the scheduled costs, the originally scheduled costs and the actual costs of the project for multiple tasks.
Baseline Report	Visio	This report can be used to display a Gantt chart of your project that is broken down by quarter and task. This report compares the originally scheduled work and costs with scheduled work and scheduled costs. Indicators show when the originally planned work exceeds the scheduled work and when the originally scheduled costs exceed the scheduled costs.

NOTES, COMMENTS:

Baseline Work Report	Excel	This report can be used to display a Gantt chart illustrating the scheduled work, the originally scheduled work and the actual work of the project for multiple tasks.
Budget Cost Report	Excel	This report can be used to display a Gantt chart illustrating the cost budget, the scheduled costs, the originally scheduled costs as well as the actual costs of the project over a specific period of time.
Earned Value Over Time Report	Excel	This report can be used to display a Gantt chart showing the actual costs for the work performed (actual costs), the scheduled value (estimated costs of the work performed) as well as the earnings value (the estimated costs of the work performed) over a specific period of time.
Budget Work Report	Excel	This report can be used to display a bar chart illustrating the budget costs of work, the scheduled work, the originally scheduled work as well as the actual work over a specific period of time.

13.2.4 CATEGORIES "TASK SUMMARY & RESOURCE SUMMARY"

Category	Type	Description
Critical Task Status Report	Visio	This report can be used to display a Gantt chart illustrating the remaining work for critical as well as non-critical tasks. This data bar indicates the percentage of the accomplished work.

NOTES, COMMENTS:

13.3 REPORTS DIRECTLY FROM MICROSOFT PROJECT

Since version 2013, Microsoft Project provides the option of directly generating reports and evaluations on the current project. Apart from preconfigured evaluations, you can also produce individual reports or customize existing reports.

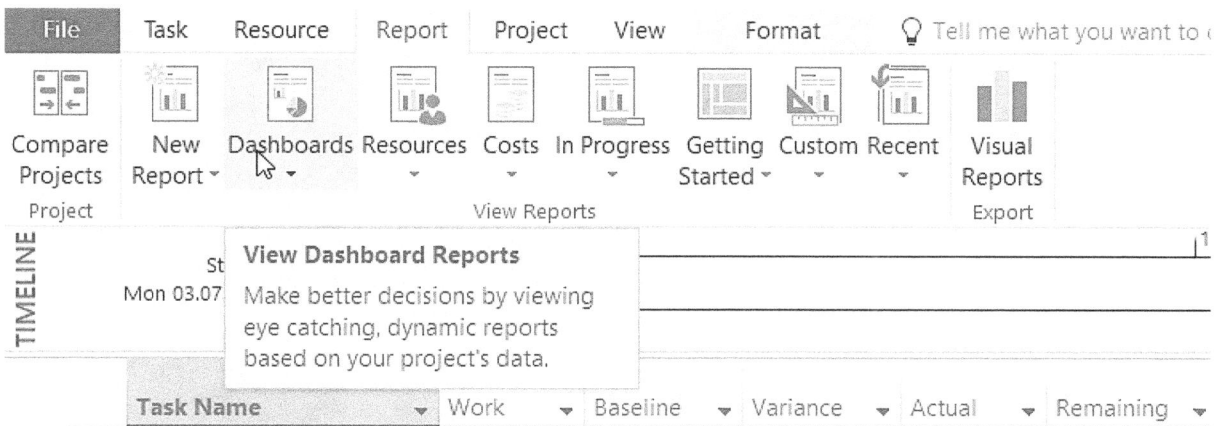

Below, you will find an example for the possible reports, a cost overview of a project with specific tasks that have already be completed or that are still in progress.

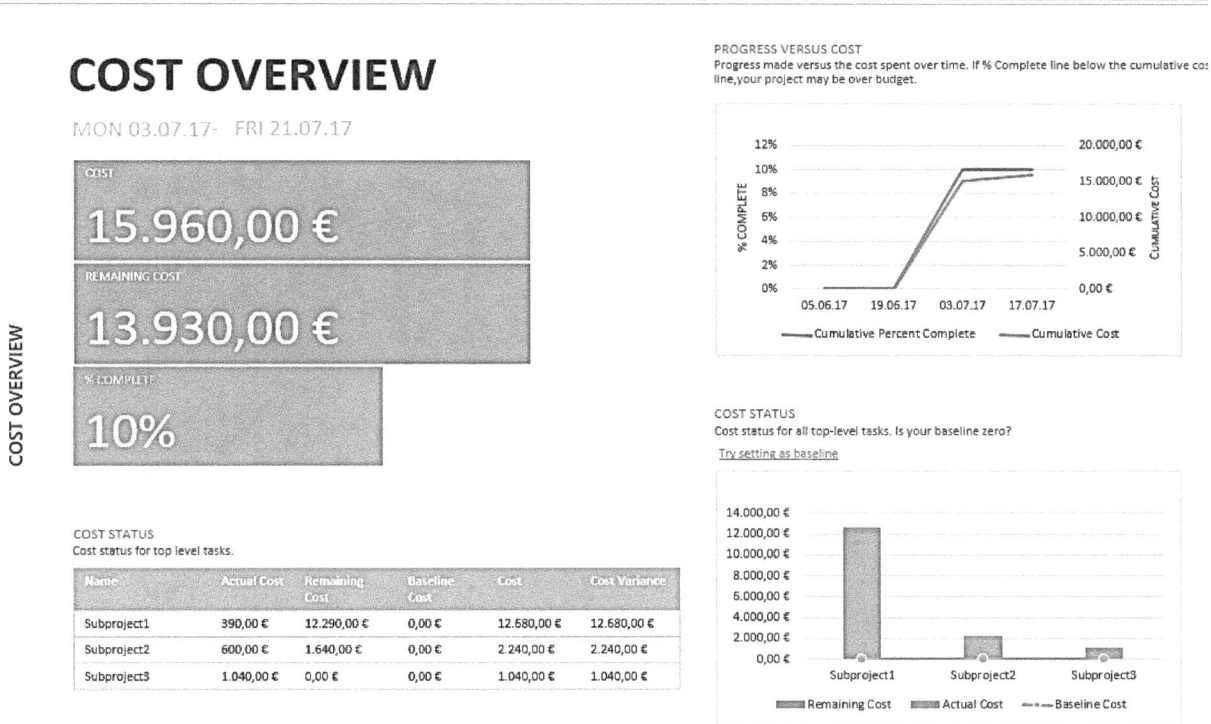

NOTES, COMMENTS:

The following standard reports are integrated and can be customized based on individual requirements.

Reports under	Content	Details on report
Dashboards	Upcoming tasks	All tasks of the current week
	Work overview	An overview of the entire scheduled work in the project
	Burndown	Graphic about the tasks to be executed
	Cost overview	Complete overview of project costs
	Project overview	Overview of the degree of completion as well as due milestones and delays
Resources	Resources (overview)	Resources scheduled in the project and status of work
	Overallocated resources	Overview report on which resources are overallocated
Costs	Earned value analysis	Report on earned value analysis
	Cost overruns	Show costs overruns compared to the baseline
	Resource costs (overview)	Overview of the resource costs
	Task costs	Cumulated costs per quarter and tasks of the first level
	Task costs (overview)	Cost overview of first-level tasks and their status
In progress	Critical tasks	Overview of critical tasks and their status
	Milestone report	Overview of the milestones in the project
	Late tasks	A list of all late tasks
	Delayed tasks	A list of all tasks that are delayed compared to the baseline

NOTES, COMMENTS:

14 ATTACHMENTS

14.1   WORKING WITH OTHER OFFICE PROGRAMS

In principle, Microsoft Project can exchange data with all other Microsoft programs, either by copy and paste or export through predefined interfaces.

Microsoft Excel:

The Microsoft Project file can be saved as an Excel workbook by selecting "Save as...". An export wizard opens displaying defined export options.

Excel especially offers dynamic links as the values in each other application automatically change as well.

Vice versa, an Excel workbook can also be opened as a Microsoft Project file.

Microsoft PowerPoint

As described in chapter "Timeline", the timeline can be copied from Microsoft Project into a PowerPoint slide and can be further edited.

Microsoft Outlook

Here, the "Timeline" from Microsoft Project can be sent from Microsoft Project as project information to any email recipient.

Mind Manager

You can export tasks from this program to Microsoft Project.

Furthermore, there are many programs that support or complement Microsoft Project in specific functions, e.g. the visual creation of project structure plans based on the outline function of Microsoft Project.

NOTES, COMMENTS:

14.2 EARNED VALUE ANALYSIS WITH MICROSOFT PROJECT

The earned value analysis (partially also referred to as earned value method or labor value analysis) is a tool of project controlling. It serves as an evaluation of the progress of projects. It describes the actual deadline and cost situation by means of indicators. The key values are planned value (PV), actual costs (AC) and earned value (EV). A trend analysis can be done by tracking the indicators.

Microsoft Project 2016 also includes the possibility to execute an earned value analysis (EVA). The associated columns/field names are listed below.

Name	Column name	Description
Planned Value (PV)	PV	Also called "Budgeted Cost of Work Scheduled" (**BCWS**). The PV is the authorized, time-phased budget assigned to accomplish the work. It's the amount that the project is supposed to be complete up to that status point.
Earned Value (EV)	EV	Also called "Budget Cost of Work Performed" (**BCWP**). The EV is the measure of the work performed at a specific point in time, expressed in terms of the approved budget authorized for that work. It's the amount that the project is actually complete up to that status point.
Actual Cost (AC)	AC	Also called "Actual Cost of Work Performed" (**ACWP**). The AC is the realized cost for the work performed during a specific time period. It's the actual cost of the work up to that status point.
Schedule Variance (SV)	SV	The schedule variance tells you how far ahead or behind schedule the task is in terms of the task budget. The formula is: $SV = EV - PV$ • If SV is negative, the task is behind schedule. • If SV is zero, the task is on schedule. • If SV is positive, the task is ahead of schedule.
Cost Variance (CV)	CV	Cost Variance (CV) is the amount that the task is over or under its budget. The formula is: $CV = EV - AC$ • If CV is negative, the task is over budget. • If CV is zero, the task is on budget. • If CV is positive, the task is under budget. For example, • CV = -$1,000 means the project is over budget. • CV = $0 means the project is right on budget. • CV = $1,000 means the project is under budget

NOTES, COMMENTS:

Estimate at Completion (EAC)	EAC	EAC is the full task or project cost expected at completion (the new project budget). There are multiple ways to calculate it based on how you expect the future of the performance of the project to be: *Future performance will be based on the budgeted cost.* If you think that the existing variance was a unique event and the rest of the project will go according to plan, simply add the remaining project budget to the actual cost incurred to date (AC). **EAC = AC + (BAC – EV)**
Estimate to Complete (ETC)	ETC	ETC represents the expected cost required to complete the project. It only measures the future budget needed to complete the project, not the entire budget. It allows the project manager to compare the funding needs to finish the project with available funding. There are two ways to calculate ETC, based on past project performance: **ETC = (BAC – EV) / CPI** and based on a new estimate. This is called "Management ETC". It means that a new estimate is created for the remaining tasks in the project.
Budget at Completion	BAC	Budget at Completion (BAC). The Budget at Completion (BAC) simply refers to the budget of each task.
Variance at Completion (VAC)	VAC	The VAC is a forecast of how the variance, in particular Cost Variance (CV), will be upon the completion of the project. It's the level of the expected cost overrun or underrun. In many situations, the project manager must request additional funding as early as possible or at least report a potential overrun. The formula is: **VAC = BAC – EAC** = old budget – new budget. It's relatively simple: If you've calculated the EAC you've already done the big math, and the 'new budget' can simply be subtracted from the 'old budget' to determine the cost overrun or underrun. "Variance at Completion" is simply a future projected "Cost Variance". It has the same units as CV and it's the same type of element.

NOTES, COMMENTS:

Cost Performance Index	CPI	Cost Performance Index (CPI), like Cost Variance, is a measure of the cost performance of the project, but it's a relative instead of an absolute measure. It tells you the cost efficiency of the project. The formula is: **CPI = EV / AC** • If CPI is less than 1, the task is over budget. • If CPI is zero, the task is on budget. • If CPI is greater than 1, the task is under budget. For example, • CPI = 0 means the project work has not started. • CPI = 0.5 means the project has spent twice the amount that it should have done at this point. • CPI = 1.0 means the project is on schedule. • CPI = 2.0 means the project has spent half the amount that it should have done at this point.
Schedule Performance Index (SPI)	SPI	Schedule Performance Index (SPI) is similar to Schedule Variance (SV). It also tells you how far ahead or behind schedule the task is in terms of the task budget, but it's a relative measure rather than an absolute one. It tells you the efficiency of the task. The formula is: **SPI = EV / PV** • If SPI is less than 1, the task is behind schedule. • If SPI is zero, the task is on schedule • If SPI is greater than 1, the task is ahead of schedule. For example, • SPI = 0 means the project work has not started. • SPI = 0.5 means the project has performed half the work it was supposed to do at this point. • SPI = 1.0 means the project is on schedule. • SPI = 2.0 means the project has performed twice the work it was supposed to do at this point. An SPI greater than 1.0 is good.

NOTES, COMMENTS:

K. Oberbörsch

To Complete Performance Index (TCPI)	TCPI	TCPI represents the efficiency level, in particular the CPI, which will finish the project on time. It can be a powerful indicator because it's generally easy to ascertain if your team will be as productive as the indicator tells you. This indicator tends to be a bigger red flag than other indicators. If it says your teams needs to be twice as efficient as the schedule requires, it tends to make you take notice that action needs to be taken: There are two ways to calculate TCPI: To achieve the <u>original budget</u> If the goal is to achieve the original project budget, i.e. the overrun or underrun has not resulted in a change to the project schedule and/or budget, the following formula applies: **TCPI = (BAC – EV) / (BAC – AC)** To achieve the <u>revised budget</u> If the goal is to achieve the project's EAC, i.e. the budget has been revised and an approved change to the project schedule/budget has occurred, use this formula. If additional funds covering the cost overrun have been requested and approved by the project sponsor, the EAC becomes the target of the project, and the following scenario applies. **TCPI = (BAC – EV) / (EAC – AC)**

NOTES, COMMENTS:

14.3 Complete List of all available Fields in Microsoft Project

Microsoft provided a detailed listing and description of all data fields under this reference. However, the current status only reflects version Microsoft Project 2013:

https://tinyurl.com/y959kz2x

14.4 Basic Settings

Timelines must be clear and look consistently to be able to merge, analyze and compare them. Therefore, Microsoft Project standards and rules are required for the design in Microsoft Project. This way, also unexperienced project managers can quickly find their way in the program and can create their own useful plans.

14.4.1 Naming Standards

- Display summary task as a substantive (example: "SPECIFICATION")

- Display milestone as a substantive + adjective (example: "Specification completed")

- Display task as a substantive + verb (example: "Create specification")

14.4.2 Standards for Summary Tasks

Summary tasks are created to outline a plan into clear, evaluable units. Subordinate tasks can be shown or hidden, and their data is totaled.

- Only allow summary tasks at the first level.

- Clearly specify the summary tasks of the first level.

- Confine yourself to a few outline levels, ideally not more than four.

- Set the upper and lower limits for the amount of tasks per summary task — e.g. at least three and not more than 20.

- Avoid links on summary tasks.

NOTES, COMMENTS:

14.4.3 STANDARDS FOR MILESTONES

Milestones are intended to mark specific tasks or points in time of the projects and to enable a project progress control. They can be divided into categories that are distinguishable and filterable by naming standards and by means of attribute fields also in the Gantt chart.

- Define a corresponding milestone for each agreed milestone.

- Distinguish between result, status, payment and handover milestones by means of attribute fields. Prefix their names to the initial letters of each milestone category (R, S, P, H,...).

- Enter the responsible persons in the field "Contact".

- Preferably, attach the date constraints only to milestones.

- In case the project includes multiple subprojects, create an independent milestone project for the key dates with external links to the subprojects. This makes much more sense than displaying the milestones in each subproject as a copy. This way, you only have to enter postponement in the central milestone plan only once and not for each subprojects separately.

- Once the project plan is approved, save the first version of the results milestones as a baseline. Don't change it anymore. Later, you can create meaningful status reports with reference to the original scheduling.

14.4.4 STANDARDS FOR TASKS

As a maximum of standardization, you can create task catalogues for repeatable projects that all include approved task descriptions. At first, this may be very time-consuming but standardizes the way to handle tasks. Notwithstanding the above, we highly recommend the following:

- Preferably, only show one resource per task.

- Depending on the project, the duration should be e.g. min. 0.5 days and max. 4 weeks.

- Put detailed information into text fields and notes instead of into task names that are too long.

- Avoid date constraints of tasks.

NOTES, COMMENTS:

14.5 BOOK RECOMMENDATIONS

Below you can find a list of books about project management that, as a writer, I find very exciting and instructive.

Paperback: 320 pages

Publisher: Dorset House Publishing Company, Incorporated; 1st U.S. Edition, 2nd Printing edition (July 1997)

Language: English

ISBN-10: 0932633390

ISBN-13: 978-0932633392

From prolific and influential consultant and author Tom DeMarco comes a project management novel that vividly illustrates the principles - and outright absurdities - that affect the productivity of a software development team.

With his trademark wit set free in the novel format, DeMarco centers the plot around the development of six software projects. Mr. Tompkins, a manager downsized from a giant telecommunications company, divides the huge staff of developers at his disposal into eighteen teams -- three for each of the software products to be built. The teams are of different sizes and use different methods, and they compete against one another and against an impossible deadline.

Managing these teams - with the help of numerous consultants who come to his aid - Mr. Tompkins tests the project management principles he has gathered over a lifetime. Each chapter closes with journal entries that make up the core of the eye-opening approach to management illustrated in this engaging novel.

Paperback: 376 pages

#1 Best Seller in Computers & Technology Industry

Publisher: IT Revolution Press; Reprint edition (October 16, 2014)

Language: English

ISBN-10: 0988262509

ISBN-13: 978-0988262508

Bill is an IT manager at Parts Unlimited. It's Tuesday morning and on his drive into the office, Bill gets a call from the CEO.

The company's new IT initiative, code named Phoenix Project, is critical to the future of Parts Unlimited, but the project is massively over budget and very late. The CEO wants Bill to report directly to him and fix the mess in ninety days or else Bill's entire department will be outsourced.

With the help of a prospective board member and his mysterious philosophy of The Three Ways, Bill starts to see that IT work has more in common with manufacturing plant work than he ever imagined. With the clock ticking, Bill must organize work flow streamline interdepartmental communications, and effectively serve the other business functions at Parts Unlimited.

In a fast-paced and entertaining style, three luminaries of the DevOps movement deliver a story that anyone who works in IT will recognize. Readers will not only learn how to improve their own IT organizations, they'll never view IT the same way again.

Paperback: 246 pages

Publisher: The North River Press;
1st edition (December 10, 2002)

Language: English

ISBN-10: 0884271536

ISBN-13: 978-0884271536

"Critical Chain," a gripping fast-paced business novel, does for Project Management what Eli Goldratt's other novels have done for Production and Marketing. Dr. Goldratt's books have transformed the thinking and actions of management throughout the world.

About the Author:

One of the world's most sought after business leaders - author and educator, Dr. Eli Goldratt. Eli Goldratt has been described by Fortune Magazine as a "guru to industry" and by Business Week as a "genius". His charismatic, stimulating, yet sometimes unconventional style has captured the attention of audiences throughout the world. Eli is a true thinker who provokes others to think.

Eli Goldratt is the creator of the Theory of Constraints (TOC) and is the author of 8 books, including the business best sellers The Goal, It's Not Luck, and Critical Chain. Goldratt's Theory of Constraints is used by thousands of companies, and is taught in hundreds of colleges, universities, and business schools. His books have sold over 3 million copies and have been translated into 23 languages. Goldratt's fascinating work as an author, educator and business pioneer has resulted in the promulgation of TOC into many facets of society and has transformed management thinking throughout the world.

14.6 GLOSSARY

Analogous technique

Creativity technique that helps finding new approaches by transferring existing approaches (e.g. from the animal world) to the current problem. The analogies can also derive from unrelated fields.

Backward pass

The second step of the network calculation in which the latest possible start and finish date of work packages is specified.

BDWP

Budgeted costs for already completed work. The field for the earning value that indicates which portion of the budget of a task should have been spent in consideration of the actual duration of the task. Microsoft Project calculates the BDWP at task and assignment level differently. See also earning value analysis.

Bottom up

Method for creating a project structure plan. At first, all tasks are collected without any order. Then they are summarized from the bottom to the top in levels and finally they are checked for completeness.

Brainstorming

Creativity technique: Production of as many spontaneous, also far-fetched, eccentric ideas in one group as possible. The ideas of other participants should also be included and further developed. Only restriction: Criticism is strictly forbidden during brainstorming.

Capacity-oriented usage planning

Scheduling in consideration of the maximum availability of the executing resources.

Capacity planning

Name and quantitative assignment of the executing capacities (resources) to each work packages required for the project in consideration of the estimated effort.

Capacity requirement (= resource requirement)

The need for staff that is required to complete all work packages of a project, identified by the estimated effort and the timing of the network.

Change management

Process that should insure that all possible types of changes of the project such as e.g. project planning or professional changes, changes of the project objective or factors that lead to a cancellation of the project, are systematically documented and thus become transparent.

Core team (= project team)

Project participants that are responsible for the project execution together with the project manager.

Critical path

All work packages of a network that can be moved in terms of time without resulting in a postponement of the project finish date are on a critical path.

Customer of a project

Overall responsible person for a plan or project. The customer approves the project budget and the framework dates.

Decision-making body

Instances of the project organization such as steering group, steering committee, controlling committee etc. Usually, they are responsible for solving cross-project conflicts and assign priorities.

Degree of completion

Percentage at which the work of a work package are completed.

Duration

Timespan from the start to the finish of a work package. Unit: Days, hours, weeks etc. It is either directly estimated or goes by the processing duration of each resource.

Effort

The effort of a work package describes the amount of work that is required to deliver a defined work result. Unit: Person days (PD), person hours (PH) etc.

Estimate effort

Estimation of the effort required to complete a work package (100% "pure project work") as well as the person in charge. It's mainly based on experiences and is the basis for capacity planning and scheduling.

Flow chart

Temporal and logical order of work packages of a project. The result of the flow chart is the network.

Finish date

Based on the flowchart, the calculated or firmly defined finish of a work package. Depending on the calculation method the following emerges:

- Earliest finish date (forward pass)
- Latest start date (backward pass)

Forward pass

The first step of the network calculation in which the earliest possible start and finish date of work packages is specified.

Free slack

The period of time by which a work package in the network can be moved without postponing another work package as well. The formula is: FF = ESD (predecessor) - FEZ (FF = Free slacks, ESD = Earliest start date, EFD= Earliest finish date)

Gantt chart

Diagram for visualizing the timing of a project. The duration of a work package is symbolized by the lengths of the bar in the timeline. The bars can contain the actual data as well as the target data. Results are displayed as points in time.

Interdisciplinary composition

Composition of a project team from employees of different segments of a company to use their different human and technical strengths to achieve the project objective.

Kick-off session (= project kick-off)

First meeting of project manager and project team for initiating a project. They discuss the project order, project objectives, contents, dates and their basic conditions. The team members are introduced to each other and they jointly determine the further strategy.

Lag (= time value)

Is assigned to a star-to-finish relationship.. It can be higher than, smaller than or equal to zero. Examples:

"Normal sequence with a lag of +3 days" means that the successor is only allowed to start 3 days after the end of the predecessor.

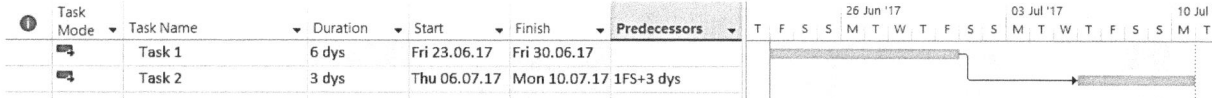

"Normal sequence with a lag of -3 days" means that the successor is already allowed to start 3 days before the end of the predecessor.

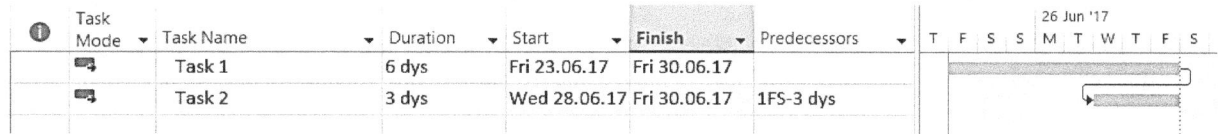

Links (= start-to-finish relationship)

Quantifiable dependency between two work packages of a project:

Normal sequence (finish-to-start)   (FS)

Start-to-start (start-to-start)   (SS)

Finish-to-finish (finish-to-finish)   (FF)

Start-to-finish (start-to-finish)   (SF)

Load diagram

Graphic for visualizing the load of employees (or divisions) by work packages from one or multiple projects.

Management style

Behavior describing the way an executive treats his employees. We distinguish between the following types

- Authoritarian management style:
  Strong focus on achieving specific objectives neglecting the social behavior of the group.

- "Laissez-faire" management style:
  Neglecting specific objectives and the social behavior of the group.

- Cooperative management style:
  Strong focus on achieving specific objectives neglecting the social behavior of the group.

Matrix project organization

Form of a project framework organization, hybrid form between a pure project organization and project coordination. Responsibility and authority are divided between project manager and all participating line functions.

Milestones

Result of particular importance during the course of the project. Usually, a milestone has a duration of = 0 days (definition for Microsoft Project).

Milestone trend analysis

Tool for controlling deadlines of a project at regular reporting dates, the scheduling of a project is re-entered visually by a query of milestone dates. A trend on meeting deadlines can be deduced from the curve progression.

Method 0/100

Method for evaluating the degree of completion of project tasks: Work packages that haven't started yet and that are still running are indicated with 0%, completed packages with 100%. This way, the so-called "90% syndrom" can be avoided.

Multi-project controlling

Analysis of the correlation of all projects to detect cross-project resource conflicts (personnel capacities, tools, finances) and be able to takes corrective measures.

Multi-project management

The task of the multiple-project management is to coordinate multiple single projects (e.g. regarding the required resources) so that the overall result of all projects concerning the corporate goals achieve an optimum.

Network

Graphic presentation of the dependencies between work packages, meaning the procedure of project execution.

Network analysis

Calculate method to determine the start and finish date of the work packages that are earliest possible and required at the latest.

Person responsible for work packages

Contact person for the project manager for the execution of work packages. The contact person doesn't necessarily have to execute all work tasks himself.

Personnel usage

Intensity that a resource needs to complete a work package. A high personnel usage results in a short processing duration and vice versa. Unit: Percentage value or person hours/day.

Phase model

Standardized project structure plan that is divided into section that are time-dependent from each other. They can occur sequentially or overlap. Example: Analysis – concept – development – realization – test

Project

Plan that fulfills the following criteria:

- Uniqueness, no routine task
- Clear target
- Temporal, financial, personnel or other constraints
- High complexity (indicators: Effort, number of participating division, risk)

Project application

A project order that has not yet been placed and that contains all information to enable you to make a decision about the usefulness of a project.

Project completion

Last phase of the project life cycle in which:

- the project result is handed over to the customer,
- the project organization is dissolved
- a summary is drawn from the past project flow (to ensure experiences gained for future projects)

After the project completion, the project is officially over.

Project completion report

Report of the project manager including a summary of the project flow

Project completion session

Last session of the project team in which the experiences from the project execution are discussed. Furthermore, it's determined how should be informed about the project completion and its result.

Project controlling (= project management)

Task of the project manager. The objective is to detect possible problems during the project execution as soon as possible to take control measures, if necessary.

Project coordination

Form a project framework organization for the duration of a project, the existing line organization is extended by the planning and control unit of a project coordinator. It doesn't have any decision or authority to issue directives towards the line functions.

Project framework organization

Interaction of project and line organization. Possible forms include:

- Pure project organization
- Project coordination
- Matrix project organization

Depending on the organizational form, the project manager has more or less responsibility and authorities.

Project phases

Sections of a project flow that are time-independent from each other. Example: Analysis – concept – development – realization – test

Project life cycle

General flow of a project from the project manager's standpoint. It contains of the following sections:

- Project start
- Project planning
- Project control
- Project completion

Project Manager

Responsible person for reaching the project objectives agreed on in the project order. He is the first contact person of the customer. Tasks, authorities and responsibility of the project manager should be specified company-wide.

Project management

Project management is a leadership concept supporting a target-oriented and efficient processing of projects. This includes organizational, methodical and interpersonal aspects.

Project management manual

This is how the documentation of basic specifications for a uniform application of project management in company is called.

Project Management Software

Helps the project manager in using planning and controlling methods, however doesn't replace the common sense.

Project organization

The project organization primarily consists of a customer, the project manager and the project team, can, however, be extended by additional control and decision committees according to the applicable requirements. The project organization will be dissolved after the end of the project.

Project participant

All persons participating in a project, even if they don't belong to the project team.

Project planning

All tasks that lead to a project plan. A project plan can consists of the following elements:

- Project structure plan incl. description of work packages
- Timeline (network, Gantt chart and milestone plan)
- Resource plan
- Cost plan
- Risk analysis

Project objective

The project objective is part of the project order and consists of three components

- Content
- Time
- Costs

It must be reachable, complete, mutually consistent, not interpretable, verifiable, solution-neutral, documented and agreed upon between customer and project manager.

Project structure

Developing a project structure plan. A project is hierarchically divided into even smaller elements. The lowest level is the basis for further project planning.

Project team (= core team)

Project participants that are responsible for the project execution together with the project manager.

Pure project organization

Form a project framework organization. For the duration of a project, the participating employees are grouped together to an independent organizational unit and assigned to the manager.

Quantitative assessment method

Method for assessing the degree of completion of project tasks: A work package is broken down into a number of similar objects with the same effort each (e.g. 30 almost similar graphs). Based on the number of the completed objects you can estimate the degree of completion. This way, the so-called "90% syndrom" can be avoided.

Reporting and information system

Generic term for the formally regulated flow of information within a project or between a project and its environment.

Resource scheduling (= usage scheduling)

Planning of the temporal usage of all resources participating in the execution of a project, depending on their availability.

Results plan

Graphic presentation of the total cost situation of a project (comparison between the estimate financial benefit and the project costs) to evaluate its profitability.

Return on Investment (ROI)

Profitability accounting to identify the return on invested capital. The ROI is calculated by the value of the generated profit divided by the value of the invested capital.

Scheduling

Planning of start and finish date of all work packages of a project. Adherence to schedules usage planning. Planning without considering the maximum availability of the executing capacities (capacity requirement planning)

Start date

Based on the flowchart, the calculated or firmly defined start of a work package. Depending on the calculation method the following emerges:

- Earliest start date (forward pass)
- Latest start date (backward pass)

Status report

Overview to be created by the project manager on the current project status (target/actual comparison of dates, costs, efforts) as an information for the customer. A status report is generated in regular intervals or when reaching specific milestones.

Step-by-step method

Method for assessing the degree of completion of work packages: A work package is divided into different sequential and temporal work steps and work steps (or steps) evaluated based on effort The degree of completion is identified by reached work steps.

Syndrome 90%

Risk of overestimating the degree of completion of a work package. The responsible person indicates to have completed a work package by 90% but the true work progress is lower.

Team 50/50

Method for evaluating the degree of completion of project tasks: Started work packages are valuated with 0%, running work packages with 50%, completed work packages with 100% degree of completion.

Top down

Method for creating a project structure plan. Based on the project target, the project is broken down in further detail by levels.

Task

In project management, a task is a clearly defined work unit that was started at a specific point in time and will be finished a specific later point in time. More generally: "a task is a process element that describes a specific event." (DIN 69900, part 1).

Usually, tasks are stages during the project flow. A task can be combined with other tasks. E.g., a task called "put on socks) would have to be finished before a task called "put on shoes" can be started.

Total slack

Period of time by which a work package in the network can be moved without having to postpone the finish date of the project.

Task Name	Start	Finish	Late Start	Late Finish	Free Slack	Total Slack
Marketing 1	Mon 03.07.17	Wed 05.07.17	Fri 07.07.17	Ved 12.07.17	0 dys	4,5 dys
Marketing 2	Thu 06.07.17	Tue 11.07.17	Ved 12.07.17	Tue 18.07.17	4,5 dys	4,5 dys
Milestone End PP	Tue 18.07.17	Tue 18.07.17	Tue 18.07.17	Tue 18.07.17	0 dys	0 dys

Usage scheduling (= resource scheduling)

Scheduling the temporal usage of all resources participating in the execution of a project, depending on their availability.

Work breakdown structure (WBS)

(Mostly graphical) overview of all required process steps for reaching the project objectives.

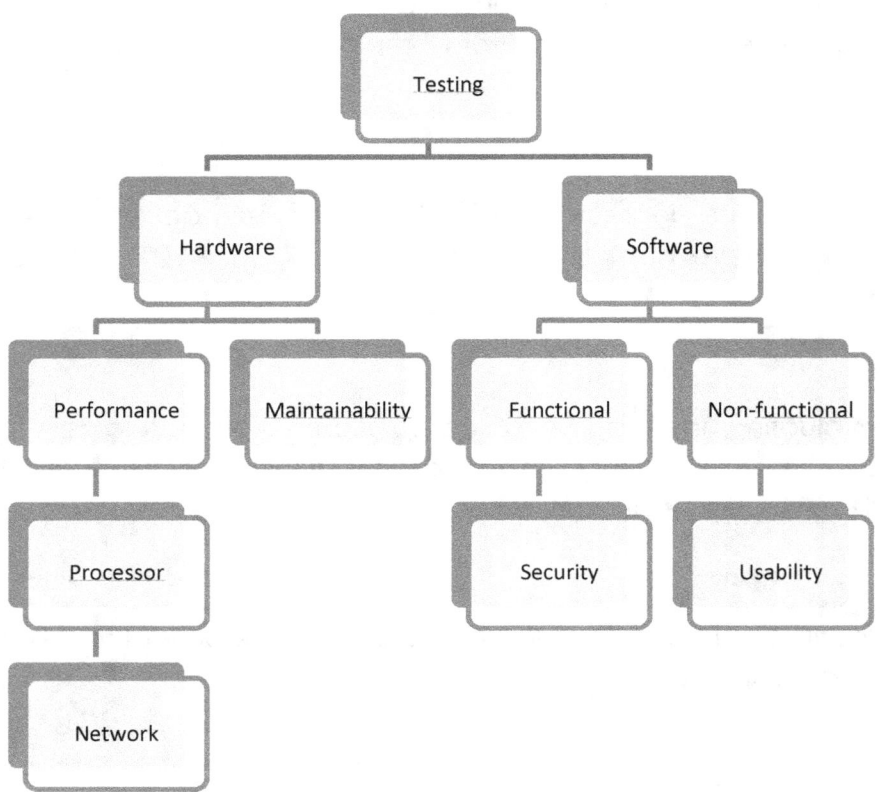

Work package

Part of a project that isn't broken down any further in the project structure plan. A work package can be located on any outline level. To reach the project objective, the completion of all work packages is required. In the current language, work packages are often also referred to as "task", "Activity" or "Process".

You can find a comprehensive glossary with approx. 1,000 terms related to project management under: https://www.projektmagazin.de/glossar/englischeliste

14.7 KEYBOARD SHORTCUTS FOR PROJECT 2016

Display and use windows

To do this	Press
Switch to the next window.	Alt+Tab
Switch to the previous window.	Alt+Shift+Tab
Close the active window.	Ctrl+W or Ctrl+F4
Restore the size of the active window after you've enlarged it.	Ctrl+F5
Move to a task pane from another pane in the program window (in clockwise direction). You may need to press F6 more than once.	F6
Move to a pane from another pane in the program window (in counterclockwise direction).	Shift+F6
If more than one window is open, switch to the next window.	Ctrl+F6
Switch to the previous window.	Ctrl+Shift+F6
Copy a picture of the screen to the clipboard.	Print Screen
Copy a picture of the selected window to the clipboard.	Alt+Print Screen

Move around in text or cells

To do this	Press
Move one character to the left.	Left Arrow
Move one character to the right.	Right Arrow
Move one line up.	Up Arrow
Move one line down.	Down Arrow

To do this	Press
Move one word to the left.	Ctrl+Left Arrow
Move one word to the right.	Ctrl+Right Arrow
Move to the end of a line.	End
Move to the beginning of a line.	Home
Move up one paragraph.	Ctrl+Up Arrow
Move down one paragraph.	Ctrl+Down Arrow
Move to the end of a text box.	Ctrl+End
Move to the beginning of a text box.	Ctrl+Home

Move around in and work in tables

To do this	Press
Move to the next cell.	Tab
Move to the preceding cell.	Shift+Tab
Move to the next row.	Down Arrow
Move to the preceding row.	Up Arrow
Insert a tab in a cell.	Ctrl+Tab
Start a new paragraph.	Enter
Add a new row at the bottom of the table.	Tab at the end of the last row

Basic file management

To do this	Press
Open a project file (display the **Open** dialog box).	Ctrl+F12
Open a project file (display the **Open** tab in the Backstage view).	Ctrl+O
Save a project file.	Ctrl+S
Create a new project.	Ctrl+N
Print a file (display the **Print** tab in the Backstage view).	Ctrl+P

Use the Open and Save As dialog boxes

To do this	Press
Display the **Open** dialog box.	Ctrl+F12
Display the **Open** tab in the Backstage view.	Ctrl+O
Display the **Save As** dialog box.	F12
Open the selected folder or file.	Enter
Open the folder one level above the open folder.	Backspace
Delete the selected folder or file.	Delete
Display a shortcut menu for a selected item such as a folder or file.	Shift+F10
Move forward through options	Tab
Move back through options	Shift+Tab
Open the **Look in** list	F4 or Alt+1

Use a Network Diagram

To do this	Press
Move to a different Network Diagram box.	Arrow keys
Add Network Diagram boxes to the selection.	Shift+Arrow keys
Move a Network Diagram box. **NOTE:** You need to set manual positioning first. Select the box you want to move. Click **Format**, and then click **Layout**. Click **Allow manual box positioning**.	Ctrl+Arrow keys
Move to the top Network Diagram box in the view or project.	Ctrl+Home or Shift+Ctrl+Home
Move to the lowest Network Diagram box in the project.	Ctrl+End or Shift+Ctrl+End
Move to the leftmost Network Diagram box in the project.	Home or Shift+Home
Move to the rightmost Network Diagram box in the project.	End or Shift+End
Move up one window height.	Page Up or Shift+Page Up
Move down one window height.	Page Down or Shift+Page Down
Move left one window width.	Ctrl+Page Up or Shift+Ctrl+Page Up
Move right one window width.	Ctrl+Page Down or Shift+Ctrl+Page Down
Select the next field in the Network Diagram box.	Enter or Tab
Select the previous field in the Network Diagram box.	Shift+Enter

Navigate views and windows

To do this	Press
Activate the entry bar to edit text in a field.	F2
Activate the menu bar.	F10 or Alt
Activate the project control menu.	Alt+Hyphen or Alt+Spacebar
Activate the split bar.	Shift+F6
Close the program window.	Alt+F4
Display all filtered tasks or all filtered resources.	F3
Display the **Field Settings** dialog box.	Alt+F3
Open a new window.	Shift+F11
Reduce a selection to a single field.	Shift+Backspace
Reset sort order to ID order and turn off grouping.	Shift+F3
Select a drawing object.	F6
Display task information.	Shift+F2
Display resource information.	Shift+F2
Display assignment information.	Shift+F2
Turn on or off the Add To Selection mode.	Shift+F8
Turn on or off Auto Calculate.	Ctrl+F9
Turn on or off the Extend Selection mode.	F8
Move left, right, up, or down to view different pages in the Print Preview window.	Alt+Arrow keys

Outline a project

To do this	Press
Hide subtasks.	Alt+Shift+Hyphen or Alt+Shift+Minus Sign (minus sign on the numeric keypad)
Indent the selected task.	Alt+Shift+Right Arrow
Show subtasks.	Alt+Shift+ = or Alt+Shift+Plus Sign (plus sign on the numeric keypad)
Show all tasks.	Alt+Shift+* (asterix on the numeric keypad)
Outdent the selected task.	Alt+Shift+Left Arrow

Select and edit in a dialog box

To do this	Press
Move between fields at the bottom of a form.	Arrow keys
Move into tables at the bottom of a form.	Alt+1 (left) or Alt+2 (right)
Move to the next task or resource.	Enter
Move to the previous task or resource.	Shift+Enter

Edit in a view

To do this	Press
Cancel an entry.	Esc
Clear or reset the selected field.	Ctrl+Delete

To do this	Press
Copy the selected data.	Ctrl+C
Cut the selected data.	Ctrl+X
Delete the selected data.	Delete
Delete row that has a selected cell.	Ctrl+Minus Sign (on the numeric keypad)
Fill down.	Ctrl+D
Display the **Find** dialog box.	Ctrl+F or Shift+F5
In the **Find** dialog box, continue to the next instance of the search results.	Shift+F4
Use the **Go To** command (**Edit** menu).	F5
Link tasks.	Ctrl+F2
Paste the copied or cut data.	Ctrl+V
Reduce the selection to one field.	Shift+Backspace
Undo the last action.	Ctrl+Z
Unlink tasks.	Ctrl+Shift+F2
Set the task to manually schedule	Ctrl+Shift+M
Set the task to auto schedule	Ctrl+Shift+A

Move in a view

To do this	Press
Move to the beginning of a project (timescale).	Alt+Home
Move to the end of a project (timescale).	Alt+End
Move the timescale left.	Alt+Left Arrow
Move the timescale right.	Alt+Right Arrow
Move to the first field in a row.	Home or Ctrl+Left Arrow
Move to the first row.	Ctrl+Up Arrow
Move to the first field of the first row.	Ctrl+Home
Move to the last field in a row.	End or Ctrl+Right Arrow
Move to the last field of the last row.	Ctrl+End
Move to the last row.	Ctrl+Down Arrow

Move in the side pane

To do this	Press
Move focus between the side pane and the view on the right side.	F6
Select different controls in the side pane if focus is in the side pane.	Tab
Select or clear check boxes and option buttons if focus is in the side pane.	Spacebar

Select in a view

To do this	Press
Extend the selection down one page.	Shift+Page Down
Extend the selection up one page.	Shift+Page Up
Extend the selection down one row.	Shift+Down Arrow
Extend the selection up one row.	Shift+Up Arrow
Extend the selection to the first field in a row.	Shift+Home
Extend the selection to the last field in a row.	Shift+End
Extend the selection to the start of the information.	Ctrl+Shift+Home
Extend the selection to the end of the information.	Ctrl+Shift+End
Extend the selection to the first row.	Ctrl+Shift+Up Arrow
Extend the selection to the last row.	Ctrl+Shift+Down Arrow
Extend the selection to the first field of the first row.	Ctrl+Shift+Home
Extend the selection to the last field of the last row.	Ctrl+Shift+End
Select all rows and columns.	Ctrl+Shift+Spacebar
Select a column.	Ctrl+Spacebar
Select a row.	Shift+Spacebar
Move within a selection down one field.	Enter
Move within a selection up one field.	Shift+Enter
Move within a selection right one field.	Tab
Move within a selection left one field.	Shift+Tab

Use a timescale

To do this	Press
Move the timescale left one page.	Alt+Page Up
Move the timescale right one page.	Alt+Page Down
Move the timescale to beginning of the project.	Alt+Home
Move the timescale to end of the project.	Alt+End
Scroll the timescale left.	Alt+Left Arrow
Scroll the timescale right.	Alt+Right Arrow
Show smaller time units.	Ctrl+ / (slash on the numeric keypad)
Show larger time units.	Ctrl+* (asterisk on the numeric keypad)

15 INDEX

A

ALLOCATING A RESOURCE.. 83
ASSIGNING RESOURCES TO TASKS.................................. 83
AUTOMATIC RESOURCE LEVELING 103
AUTOMATIC SCHEDULING... 29

B

BAR STYLES... 19
BASIC INFORMATION.. 23
BASIC SETTINGS 185
BOOK RECOMMENDATIONS 189
BUDGET TRACKING.. 123
built-in/default tables 69

C

CALENDAR 25
CAPTURING TARGET (SET BASELINE)........................... 135
CELLS HIGHLIGHTNING 133
CHARACTERISTICS OF LINKING..................................... 39
CODE 79
COMAPRE TARGET VALUES AGAINST ACTUAL VALUES IN A
    TABLE.. 137
COMPARE TARGET AND ACTUAL VALUES VISUALLY 139
COST MANAGMENT... 111
COST TYPES 111
COSTS PER TASK 77
COSTS/USE 79
CRITICAL PATH 61
CRITICAL TASKS... 63
CURRENT DATE 23
CUSTOM FIELDS 25
CUSTOM FIELDS 147
CUSTOMIZED ADAPTIONS 71

D

DATA FIELDS 69
DEACTIVATING "EFFORT-DRIVEN"................................. 97
DEPENDENCIES 35
DIAGRAM AREA 11
DIFFERENT RESOURCE TYPES 77
DISPLAY OF TOOLBARS AND TABS 13
DISPLAY RESOURCE USAGE 85
DURATION OF A TASK 27

E

ENTERING TASKS.. 33
EVALUATING MONITORING INFORMATION.................... 145

F

FILE TAB... 13
FILTER FUNCTIONS 129
FINISH DATE 23
FINISH-TO-FINISH (FF).. 35
FINISH-TO-START (FS)... 35
FIXED DURATION, EFFORT-DRIVEN 95
FIXED UNITS, EFFORT-DRIVEN 89
FIXED UNITS, NON-EFFORT-DRIVEN.............................. 93
FIXED WORK, EFFORT DRIVEN 99
FONTS AND FORMAT 15
FORMAT TAB 19

G

GANTT CHART STYLE 19
GLOSSARY 192
GROUP 79
GROUPING 131

I

INDEX 213
INDICATOR COLUMN 51, 53
INITIALS 79
INSERT SUBPROJECT 17
INTRODUCTION... 5

K

KEYBOARD SHORTCUTS.. 203

L

LEVELING OPTIONS... 15, 103
LOOKUP FIELDS 147

M

MANUAL RESOURCE LEVELING 105
MANUAL SCHEDULING.. 29
MATERIAL LABEL 79
MAX. UNITS 79
MILESTONES 15, 33, 55, 131, 187, 196
MORE TABLES 71
MULTIPLE TIMELINES... 59
MULTI-PROJECT MANAGEMENT 153

N

NETWORK DIAGRAM 61

O

ONE-TIME COSTS 117
ONGOING DURATIONS.. 27
ONLINE TEMPLATES 21
OUTLINE LEVELS 43
OUTLINE NUMBER 45
OVERALLOCATED 85
OVERTIME RATE 79
OVERVIEW OF PROGRAM STRUCTURE AND DESIGN.......... 11
OVERVIEW OF VISUAL REPORTS 165

P

PLANNED BUDGET 127
PLANNING STEPS 9
POST-IT NOTE 53
PREDECESSOR...37, 39
PREDEFINED REPORTS.. 145
PREDETERMINED TEMPLATES 21
PRODUCT PORTFOLIO 153
PROJECT BASED RESOURCES.. 75
PROJECT CONTINUATION 143
PROJECT CONTROL/MONITORING 135
PROJECT INFORMATION 17
PROJECT TAB.. 17
PROJECT/SUBPROJECT 155
PROJECT/TASK VIEWS 129
PROJECTPORTFOLIO/OVERVIEW.. 161
PSP CODE 45

Q

QUICK ACCESS TOOLBAR 11

R

REPORT TAB 17
REPORT/VISUAL REPORTS... 119
REQUIRED RESOURCES 73
RESOURCE INFORMATION 79
RESOURCE LEVELING.. 103
RESOURCE PERSPECTIVE.. 85
RESOURCE POOL...................73, 75, 153, 157, 159, 161
RESOURCE: TABLE 75
RESSOURCE TAB 15

S

SAVING BASELINE.. 135
SCHEDULE FROM.. 23
SCHEDULING FORMULA.. 99
SCHEDULING RESOURCE USAGE 73
SCHEDULING RESOURCES... 73

SCHEDULING TOPDOWN SUMMARY TASK 47
SECOND TIMELINE 59
SETTING UP A NEW PROJECT 21
SEVERAL TIMELINES.. 7
SLACK TIMES 65
SPECIFYING A DEADLINE 49
STANDARD CALENDAR 25
STANDARD RATE.. 79
STANDARD REPORTS 173
STANDARD TABLES 69
START DATE.. 23
START-TO-FINISH (SF) 35
START-TO-START (SS) 35
STATUS 17
STATUS BAR 31
STATUS DATE 23
STRUCTURING TASKS 41
SUBTASKS 41, 47
SUCCESSOR...35, 37
SUMMARY TASK 41

T

TABLE 11
TABLES 69
TASK CONSTRAINTS 49
TASK MODES AND EFFORT TRACKING 87
TASK NOTES 53
TASK SCHEDULING 27
TASK TAB 15
TASKS MODE 15
TEAM PLANNER 15, 19, 101
TELL ME WHAT YOU WANT TO DO? 7
TEXT STYLES... 19
TIMELINE... 57
TIMELINE BAR 59
TOTAL SLACK 65
TRAFFIC LIGHTS 123
TRAVEL EXPENSES... 115

V

VIEW TAB.. 19
VISUAL REPORTS 17
VISUAL REPORTS... 163

W

WORKING WITH OTHER OFFICE PROGRAMS 175

Z

ZOOM.. 19

Author:

KLAUS OBERBÖRSCH, born 1955

Over 30 years of IT experience in various industries and an overall professional experience of 45 years. Introduced MS Project Server, e.g. at SAP, FinanzIT. Has worked as a trainer for Microsoft Project for more than 25 years.

Achieved certifications in project management: GPM-IPMA - Project Management Expert, Prince2, ASQF Certified Professional for Project Management

Received certifications in IT, software engineering and testing: ISTQB Certified Tester Foundation Level, ISTQB Advanced Level Test Manager, Expert Merchant for Data Processing Organization and Data Communication (Chamber of Industry and Commerce IHK), Insurance Professional (Chamber of Industry and Commerce IHK)

Currently working as an accredited senior executive trainer conducting certification courses such as ISTQB Certified Tester Foundation Level, ISTQB Advanced Level Test Manager and ASQF Project Management with a pass rate above average across Europe.

Thanks to my wife Karin who has critically reviewed the document and allowed me to spend the required time to write this book.

Translator:

BRITTA WEBER – Translation Weber – www.translation-weber.de

Extensive experiences in translating all types of technical and marketing texts mainly for the IT and engineering industry (website content, guidebooks, presentations, technical documentations, product sheets, press releases, manuals, training materials) into English and German. Has worked with renowned medium-size companies, large enterprises (the big players), IT freelancers, start-ups and marketing/advertisement agencies.

Over the years, she has gained profound know-how in different areas of IT, such as professional software, Cloud computing, IoT, digitalization, virtualization, server technology etc. Apart from her personal passion for IT and its constant changes, this is the basis for truly precise and accurate translations to the point.

Translation Weber supports IT customers in improving their internal and external communication and thus helps them in strengthening their market position and competitiveness.